JN292762

Material Circulation through Agro-Ecosystems in East Asia and Assessment of its Environmental Impact

**Edited by
Yousay Hayashi**

Proceedings of an International Workshop on
Material Circulation through Agro-Ecosystems in
East Asia and Assessment of its Environmental Impact
– Toward International Collaboration Research –
Tsukuba, Japan
March 25-27, 2003

National Institute for Agro-Environmental Sciences
Tsukuba, Japan

NIAES Series 5

Organizing Committee

Chairman:
MINAMI, K., Director General, NIAES, Japan
Members:
China
ZHOU, J., Director General, Institute of Soil Science, CAS, China
YANG, L., Deputy Director General, Institute of Soil Science, CAS, China
Korea
EOM, K., Director, National Institute of Agricultural Science and Technology, RDA, Korea
KOH, M., Deptuy Director, National Institute of Agricultural Science and Technology, RDA, Korea
Japan
HAYASHI, Y., Director, Department of Global Resources
OKA, M., Director, Department of Biological Safety
IMAI, H., Director, Department of Environmental Chemistry
ISHII, Y., Director, Chemical Analysis Laboratory
UWASAWA, M., Director, Natural Resources Inventory Center
SEINO, H., Director, Department of Research Planning and Coordination
IMAGAWA, T., Head, Division of Research Planning, Department of Research Planning and Coordination

ISBN4-8425-0368-8 C3061

All rights reserved.
No part of this book may be translated or reproduced in any from without written permission from NIAES.

© 2004 by NIAES
Printed in Japan by YOKENDO Publishers, Tokyo

COVER PHOTO
© SCIENCE PHOTO LIBRARY ／amana

Contributors

Cai, Zucong
   Institute of Soil Science, Chinese Academy of Sciences, Nanjing 210008, China

Coleman, Kevin
   IACR-Rothamsted, Harpenden, Herts AL5 2JQ, UK

Eom, Ki-Cheol
   National Institute of Agricultural Science and Technology, Suwon, 441-707, Korea

Gong, Zitong
   Institute of Soil Science, Chinese Academy of Sciences, Nanjing 210008, China

Hayashi, Yousay
   National Institute for Agro-Environmental Sciences,
   3-1-3 Kannondai, Tsukuba, Ibaraki 305-8604, Japan

Hua, Xu
   Institute of Soil Science, Chinese Academy of Sciences, Nanjing 210008, China

Hur, Seung-Ho
   National Institute of Agricultural Science and Technology, Suwon 441-707, Korea

Itahashi, Sunao
   National Institute for Agro-Environmental Sciences,
   3-1-3 Kannondai, Tsukuba, Ibaraki 305-8604, Japan

Jung, Goo-Bok
   National Institute of Agricultural Science and Technology, Suwon 441-707, Korea

Jung, Pil-Kyun
   National Institute of Agricultural Science and Technology, Suwon 441-707, Korea

Kang, Kee-Kyung
   National Institute of Agricultural Science and Technology, Suwon 441-707, Korea

Kim, Tae-Wan
   College of Agriculture and Life Sciences, Hankyong National University, Ansung, Korea

Kobara, Yuso
   National Institute for Agro-Environmental Sciences,
   3-1-3 Kannondai, Tsukuba, Ibaraki 305-8604, Japan

Kobayashi, Kazuhiko
    Graduate School of Agricultural and Life Sciences, The University of Tokyo,
    1-1-1 Yayoi, Bunkyo-ku, Tokyo 113-8657, JAPAN

Koh, Mun-Hwan
    National Institute of Agricultural Science and Technology, Suwon 441-707, Korea

Makino, Tomoyuki
    National Institute for Agro-Environmental Sciences,
    3-1-3 Kannondai, Tsukuba, Ibaraki 305-8604, Japan

Minami, Katsuyuki

    National Institute for Agro-Environmental Sciences,
    Kannondai 3-1-3, Tsukuba, Ibaraki 305-8604, Japan

Mochizuki, Atsushi
    National Institute for Agro-Environmental Sciences,
    3-1-3 Kannondai, Tsukuba, Ibaraki 305-8604, Japan

Nakai, Makoto
    National Institute for Agro-Environmental Sciences,
    3-1-3 Kannondai, Tsukuba, Ibaraki 305-8604, Japan.

Nakatani, Yukinobu
    National Institute for Agro-Environmental Sciences,
    3-1-3 Kannondai, Tsukuba, Ibaraki 305-8604, Japan

Nam, Jae-Jak
    National Institute of Agricultural Science and Technology, Suwon 441-707, Korea

Obara, Hiroshi
    National Institute for Agro-Environmental Sciences,
    3-1-3 Kannondai, Tsukuba, Ibaraki 305-8604, Japan.

Ohkura, Toshiaki
    National Institute for Agro-Environmental Sciences,
    3-1-3 Kannondai, Tsukuba, Ibaraki 305-8604, Japan.

Otani, Takashi
    National Institute for Agro-Environmental Sciences,
    3-1-3 Kannondai, Tsukuba, Ibaraki 305-8604, Japan

Pan, Xianzhang
    Institute of Soil Science, Chinese Academy of Sciences, Nanjing 210008, China

Park, Back-Kyun
    National Institute of Agricultural Science and Technology, Suwon 441-707, Korea

Park, Chang-Young
    National Yeongnam Agricultural Experiment Station, Milyang 627-130, Korea

Petersen, Gary.W.
    Environmental Resources Research Institute, The Pennsylvania State University, University Park, USA

Sakurai, Yasuhiro
    National Institute for Agro-Environmental Sciences,
    3-1-3 Kannondai, Tsukuba, Ibaraki 305-8604, Japan

Seike, Nobuyasu
    National Institute for Agro-Environmental Sciences,
    3-1-3 Kannondai, Tsukuba, Ibaraki 305-8604, Japan

Seo, Myung Chul
    National Institute of Agricultural Science and Technology, Suwon 441-707, Korea

Shi, Weiming
    Institute of Soil Science, Chinese Academy of Sciences, Nanjing 210008, China

Shi, Xuezheng
    Institute of Soil Science, Chinese Academy of Sciences, Nanjing 210008, China

Shirato, Yasuhito
    National Institute for Agro-Environmental Sciences,
    Kannondai 3-1-3, Tsukuba, Ibaraki 305-8604, Japan

Sho, Kyu-Ho
    National Institute of Agricultural Science and Technology, Suwon 441-707, Korea

Sugahara, Kazuo
    National Institute for Agro-Environmental Sciences,
    3-1-3 Kannondai, Tsukuba, Ibaraki 305-8604, Japan

Sun, Weixa
    Institute of Soil Science, Chinese Academy of Sciences, Nanjing 210008, China

Tadano, Toshiaki
    Faculty of Agriculture, Hokkaido University, Sapporo 060-8589, Japan

Takeuchi, Makoto
    National Institute for Agro-Environmental Sciences,
    3-1-3 Kan-nondai, Tsukuba, Ibaraki 305-8604, Japan

Togami, Kazuki
    National Institute for Agro-Environmental Sciences,
    3-1-3 Kannondai, Tsukuba, Ibaraki 305-8604, Japan

Warner, Eric
    Environmental Resources Research Institute, The Pennsylvania State University, University Park, USA

Yan, Weidong
    Institute of Soil Science, Chinese Academy of Sciences, Nanjing 210008, China

Yang, Changming
    Institute of Soil Science, Chinese Academy of Sciences, Nanjing 210008, China

Yang, Jae E.
    College of Agriculture and Life Sciences, Kangwon National University, Chunchon, Korea

Yang, Linzhang
    Institute of Soil Science, Chinese Academy of Sciences, Nanjing 210008, China

Yasuda, Koji
    National Institute for Agro-Environmental Sciences,
    3-1-3 Kannondai, Tsukuba, Ibaraki 305-8604, Japan

Yokozawa, Masayuki
    National Institute for Agro-Environmental Sciences,
    Kannondai 3-1-3, Tsukuba, Ibaraki 305-8604, Japan

Yoshimatsu, Shin-ichi
    National Institute for Agro-Environmental Sciences,
    3-1-3 Kannondai, Tsukuba, Ibaraki 305-8604, Japan

Yu, Dongsheng
    Institute of Soil Science, Chinese Academy of Sciences, Nanjing 210008, China

Zhang, Yong-Seon
    National Institute of Agricultural Science and Technology, Suwon 441-707, Korea

Preface

Katsuyuki Minami

Director General, National Institute for Agro-Environmental Sciences (NIAES)

A vast variety of materials circulate through the atmosphere, hydrosphere, pedosphere, and biosphere. From the mid-20th century onward, however, unrestrained human activities have been disturbing the sustainability of global material circulation, and brought about various imbalances in material flows, resulting in a rise in greenhouse gases (GHG), water shortages, soil deterioration, natural resource exhaustion, accumulation of toxic chemical substances, and a reduction in biological diversity.

The Earth Summit 1992 sounded an alarm bell heard all over the world, calling attention to the importance of global-scale environmental conservation for human existence. To cope with the current worrying situation, a series of international pacts concerning global warming have been concluded among the member nations of the Intergovernmental Panel on Climate Change (IPCC), and the Kyoto Protocol. In addition, great efforts have been made internationally in preserving endangered biological species, establishing international standards for toxic chemical pollutants, such as dioxins and cadmium. Consequently, the Convention on Biological Diversity (CBD) was also concluded.

East Asian countries, in particular, China, Korea, and Japan, share many agro-environmental problems in common, such as the emission of GHG from farmland, effects of elevated CO_2 on crop growth, pollution of soils and crops by heavy metals and dioxins, deterioration of water quality, and the environmental risks of invasive species and GMO & GM-crops. The National Institute for Agro-Environmental Sciences, the leading institute of agro-environmental research in Japan, has successively concluded Memorandums of Understanding, in 2001 with the National Institute of Agricultural Sciences and Technology of the Rural Development Administration, Republic of Korea, and in 2002 with the Institute of Soil Science of the Chinese Academy of Sciences, the People's Republic of China, in order to enhance the three-party research collaboration on shared agro-environmental problems, and to work out better solutions collaboratively.

As the first high-profile step to implement research collaboration among the three parties, at the International Workshop on Material Circulation, the top scientists from China, Korea, and Japan reported relevant information regarding agro-environmental problems in East Asia. In this workshop, scientists from the three institutes discussed important issues common to agro-environments in East Asia, to aim for feasible and effective schemes of collaborative research in order to work out better solutions to correct the poorly balanced material cycle in agro-environments, and reduce its environmental impact.

This monograph is a compilation of research papers presented at the international workshop, which was organized by NIAES. I am confident that this monograph will serve as a valuable reference for researchers, students, and extension staff in their efforts to develop more effective approaches to material circulation through agro-ecosystems.

July 2004

Contents

Contributors ··· i
Preface ··· v

Keynote address

Global nitrogen and carbon cycles, agriculture-related disruption of these cycles,
   and the environmental consequences
      K. Minami ·· 1

Part 1. Regional Agriculture and Environment in East Asia

Environmental and ecological aspects of Korean agriculture
      K. C. Eom, T. W. Kim and J. E. Yang ···························· 17
Agro-climatological backgrounds for impact assessment in East Asia and speculation
   about future resources
      Y. Hayashi ·· 35

Part 2. Greenhouse Gas Emission and Sequestration in Agro-Ecosystem

Options for mitigating CH_4 emissions from rice fields in China
      Z. Cai and X. Hua ·· 45
FACE : a window through which to observe the responses of agricultural ecosystems to
   future atmospheric conditions
      K. Kobayashi ·· 57
Estimating carbon sequestration in Japanese arable soils using the Rothamsted carbon model
      Y. Shirato and M. Yokozawa ···································· 69

Part 3. Cycling of Farm Chemicals in Agro-Ecosystem

Studies on the coordination of the soil phosphorus supply with rice photosynthesis
   under different nutrient treatments
      C. Yang, L. Yang, T. Tadano ···································· 77
Development of a comprehensive system for the analysis and evaluation of water quality
   in medium-sized watersheds
      S. Itahashi and M. Takeuchi ···································· 89

Distribution and long-term changes in levels of polycyclic aromatic hydrocarbons in South Korean soils
  J. J. Nam, B. K. Park, K. H. Sho and C. Y. Park ············ 103

Temporal changes in dioxin levels in Japanese paddy fields
  N. Seike and T. Otani ············ 119

Development of a crop-soil database for evaluation of the risk of cadmium contamination in staple crops
  K. Sugahara, T. Makino and Y. Sakurai ············ 129

Estimating methyl bromide emissions due to soil fumigation, and techniques for reducing emissions
  Y. Kobara ············ 137

Development of technologies for assessing nutrient losses in agricultural ecosystems in Korea
  P. K. Jung, Y. S. Zhang, S. H. Hur, K. K. Kang and M. C. Seo ············ 151

Part 4. GMO, Bio-Remediation and Bio-Diversity

Components of root exudates from genetically modified cotton
  W. Shi and W. Yan ············ 157

Methods for assessing the indirect effects of introduced hymenopteran parasitoids on Japanese agricultural ecosystems
  A. Mochizuki ············ 167

Part 5. Construction of Environmental Resources Inventory and Its Utilization

Monitoring of environmental resources and their utilization in South Korea
  M. H. Koh, G. B. Jung and K. C. Eom ············ 175

A 1:1 000 000-scale soil database and reference system for the People's Republic of China
  X. Shi, D. Yu, X. Pan, W. Sun, Z. Gong, E. D. Warner and G. W. Petersen
············ 193

Construction and use of a soil inventory by NIAES, Japan
  H. Obara, T. Ohkura, K. Togami and M. Nakai ············ 199

Construction of an insect inventory and its utilization by Japan's NIAES
  K. Yasuda, S. Yoshimatsu and Y. Nakatani ············ 207

Keynote address

Global nitrogen and carbon cycles, agriculture-related disruption of these cycles, and the environmental consequences

Katsuyuki Minami

National Institute for Agro-Environmental Sciences, 3-1-3 Kannondai, Tsukuba, Ibaraki 305-8604, Japan

Introduction

The Japanese writer Ryotaro Shiba wrote the essay *To You Who Live in the 21st Century* (my translation of the Japanese title) for a Japanese textbook for sixth-graders. The following is an excerpt:

"One thing has not changed since far in the past and will not change far into the future: the elements of nature such as air, water, and soil have existed and will continue to exist, and humans, other animals, plants, and microorganisms all depend on these elements for their survival. Nature is the one constant. Humans cannot survive without air to inhale and will die from desiccation without water... In the 21st century, science and technology will continue to advance, like a flood, but will not swallow humanity. I hope that you will control this flood and put it on the right track, just as you would correct the flow of a river that threatens your village."

In one sense, Shiba is correct in saying that nature is a *constant*. However, is it true that nature is *unchanging*? Unfortunately, no. The air is not what it used to be. The atmospheric concentrations of substances that promote global warming, such as carbon dioxide, methane, and nitrous oxides, have been rapidly increasing. Substances such as chlorofluorocarbons (CFCs), nitrous oxides, and methyl bromide (CH_3Br), which deplete the ozone layer, have been increasing in the atmosphere. Acid rain falls from the air. Water is also not what it used to be. Much groundwater contains high levels of nitrate and is unsuitable for drinking. River water has undergone eutrophication in response to increased nitrogen and phosphorus concentrations. Soils have suffered similar fates. Soils that have been strenuously nurtured over tens of thousands of years are washed into rivers in an instant by erosion caused by clearing. Many soil microorganisms are dying because of excessive applications of agricultural chemicals and fertilizers. We can almost hear the Earth's shrieks of pain all over the world.

Why is the "one constant" changing ? It is because the science and technology we have developed during the 20th century are about to swallow nature like a flood. In other words, our science and technology have changed natural cycles at a global level. If science and technology had been developed and used based upon ecosystem principles, the "one constant" would indeed have remained unchanged.

The increasing abnormality of the world's cycles has become a global environmental problem, starting with point pollution by heavy metals such as cadmium and mercury and developing into more widespread problems that cover extensive areas, such as the eutrophication of lakes by nitrogen and phosphorus. Similarly wide-ranging problems are caused by emissions of various chemical substances into the atmosphere as a result of human activities and population increases. Global environmental changes such as global warming and acid rain are only two examples.

Many global environmental problems are closely related to agricultural production, including global warming, depletion of the ozone layer, acid rain, and soil erosion. Simultaneously, pressures are mounting on agriculture, and we currently face serious issues concerning food security and environmental conservation. What is more serious is the growing concern about the interactions between various forms of environmental deterioration. Many of these problems are linked with the basic nature of global nitrogen and carbon cycles, among others.

Interactions between various phenomena that damage the environment, including the interaction between global warming and ultraviolet rays, are critical determinants of environmental deterioration at the global level. In this paper, I identify alterations of the nitrogen and carbon cycles and interactions between impacts related to agriculture, and discuss the future implementation of technical and other measures that could mitigate or resolve these problems.

1. Everything flows

Nitrogen and carbon: these elements illustrate the importance of the observations that "everything flows" and the principle that "all things are in flux" better than any other elements. Humans have created many substances such as plastics, radioactive materials, and chlorofluorocarbons that cannot "flow". Instead, they are disposed of without being allowed to enter the "flow", continue to accumulate, and are driving the Earth, where we live, to extremity. If we do not resign their souls to a "flowing system" ? the true nature of nitrogen and carbon — the damage to the Earth will be irreparable some day.

However, we humans have already seriously interrupted the flow of nitrogen and carbon. The biggest effects on the environment have come from rising atmospheric concentrations of nitrous oxide (N_2O), carbon dioxide (CO_2), and methane (CH_4) and rising concentrations of nitrate (NO_3^-) in groundwater. These increasing nitrogen and carbon concentrations, combined with changes in their cycles in the biosphere, have caused various changes in the global environment, including global warming, destruction of parts of the ozone layer, acid rain, and extensive water pollution. As a result, all areas of the biosphere from the groundwater to the stratosphere are now exposed to the threat of global environmental change.

For nitrogen, such changes originated with the artificial fixation of atmospheric nitrogen (N_2, 78% percent of the air) by humans at a pace faster than the natural rate of nitrogen fixation by living organisms. In addition, this artificial nitrogen fixation at the Earth's surface has been done without any consideration of factors such as balance within ecological systems, the location of the fixation, and the time. As a result, nitrogen cycles have been changed for the worse.

A similar phenomenon can be observed for carbon. Carbon fixed at the Earth's surface and in the planet's crust is released into the atmosphere in oxidized form by burning carbon sources such as coal and oil products. As a result, atmospheric concentrations of carbon dioxide have continued to rise. Reduced carbon in the form of methane is created by the metabolism of microorganisms, and is often stimulated by human activities such as livestock production and the flooding of rice paddies. An increase in the concentrations of both gases in the atmosphere is causing global warming.

For the smooth cycles of nitrogen and carbon to be restored, it is necessary to fix the conditions that sustain these cycles and to control and restrict (where necessary) the "flows" of both natural and artificial substances.

2. Changes in the forms of nitrogen and carbon and in their cycles

In this section, I summarize changes in the forms of nitrogen and carbon and in their cycles in the anthroposphere.

2.1 Nitrogen

Nitrogen is essential for life. It is a fundamental component of the amino acids that are indispensable to the maintenance of life. Many amino acids are combined to form proteins, which act as enzymes or store small molecules such as oxygen. Moreover, nitrogen is an essential component of the bases that make up DNA, the molecule that carries the genetic codes of all higher organisms, and RNA, the molecule that allows DNA to guide the construction of the cells that make up these organisms and control the activity of these cells once they have been built. (RNA is also the genetic code for most lower organisms.) Nitrogen plays many other important biological roles. For example, many organisms use nitrogen in an oxidative form as a substitute for oxygen during respiration. Conversely, others oxidize reduced nitrogen by combining it with oxygen and thereby liberate energy. Perhaps most importantly, plant growth is not possible without nitrogen in either oxidized or reduced form, and all higher organisms depend on plants for their survival, both as a source of oxygen in the air and as source of sustenance.

Nitrogen has various important aspects when oxidized and reduced (Fig. 1). For example, its oxidation states range between +5 and − 3, and nitrogen compounds in these states that are important in the ecosystem include NO_3^-, NO_2^-, NO_2, NO, N_2O, N_2, NH_3, and NH_4^+. This means that biologically mediated redox reactions control significant parts of the nitrogen cycle. The fact that the combinations of nitrogen change in accordance with oxidation and reduction reactions shows how nitrogen moves from one reservoir to another in different

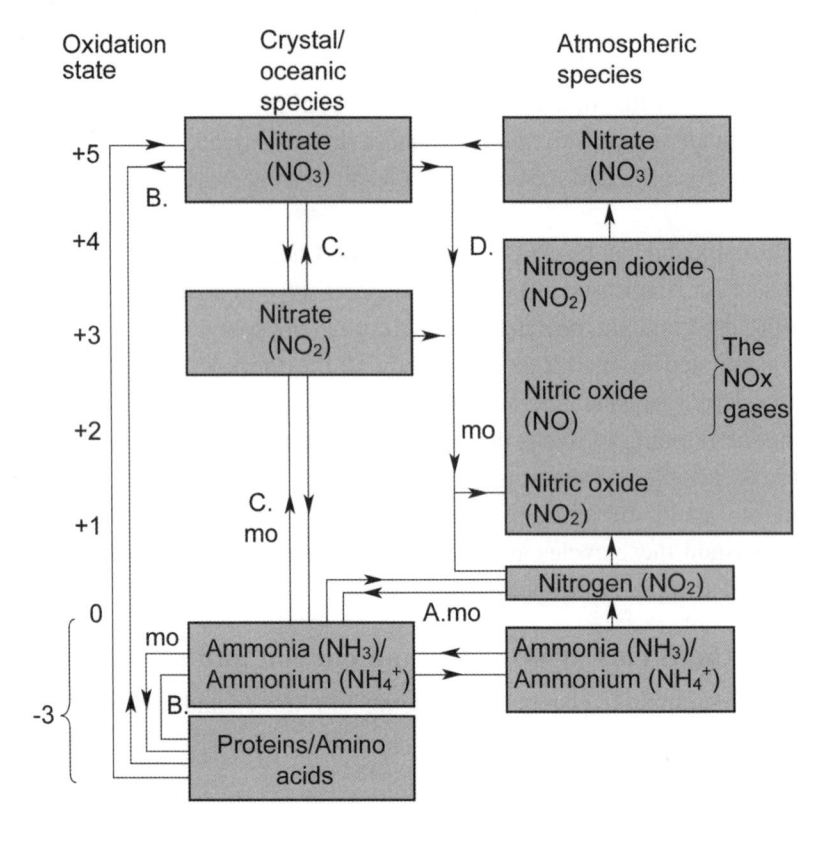

Fig. 1 Nitrogen speciation in the environment (Jackson and Jackson, 1996).

forms and at different physical locations.

Nitrogen fixation, denitrification, and nitrification involve the transfer of nitrogen from one system to another. Nitrogen fixation is a reaction that creates another nitrogen compound from molecular nitrogen gas (N_2) in the atmosphere. Nitrogen fixation is performed both biologically and industrially. In industrial nitrogen fixation, nitrogen gases are reduced to ammonia by means of the Haber-Bosch process. However, this process requires high pressure and high temperature, thereby consuming a great amount of energy. On the other hand, biological fixation is performed under natural environmental conditions by nitrogen-fixing organisms in the soil. In these biological processes, nitrogen gas is reduced to ammonia by nitrogenase enzymes, then is absorbed by organisms. This process is a key part of global nitrogen cycles because it creates a source of nitrogen for plant nutrition.

Denitrification is mainly a process performed by anaerobic bacteria, in the absence of

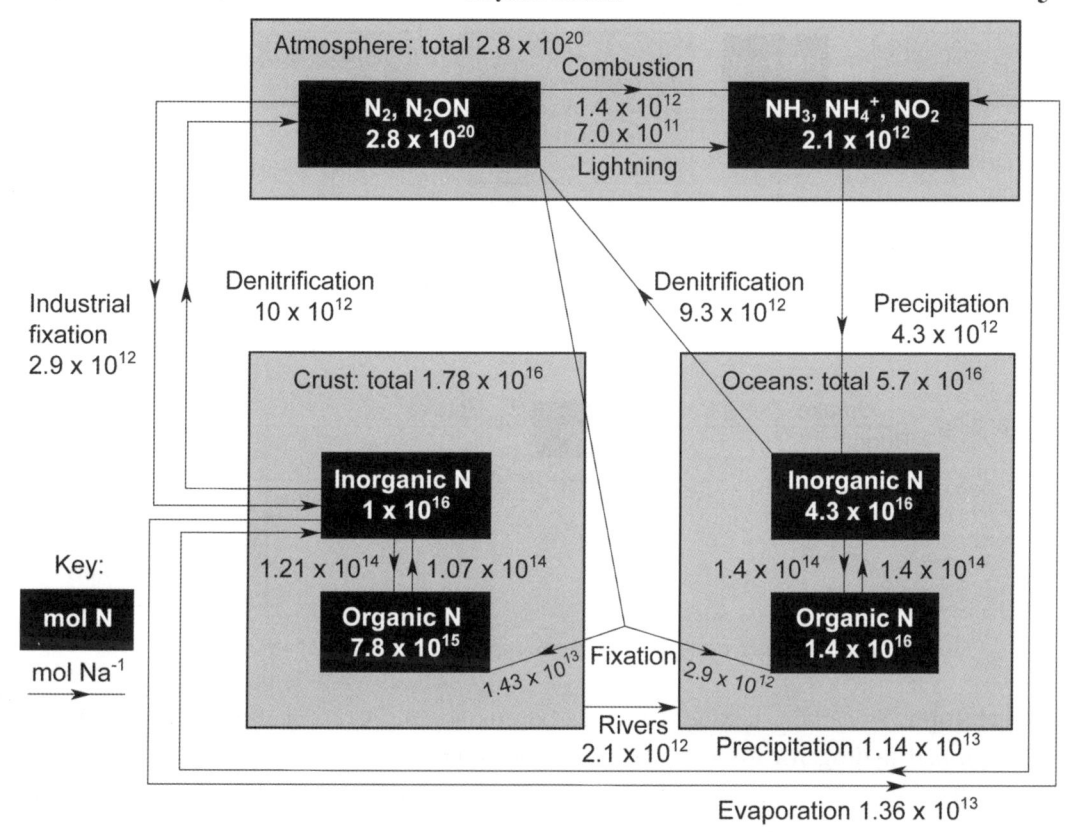

Fig. 2 The nitrogen cycle (Jackson and Jackson, 1996).

oxygen, and produces N_2 or N_2O by using nitrite and nitrate nitrogen (NO_2^- and NO_3^-) to accept electrons generated by respiration. Conversely, nitrification is a reaction that involves of the oxidation of ammonia nitrogen by microorganisms and the production of nitrite and nitrate nitrogen. Nitrification is performed by nitrifying bacteria that are ubiquitous in the soil. Nitrifying bacteria are divided into two main types — ammonium-oxidizing bacteria that oxidize ammonia nitrogen into nitrite nitrogen, and nitrite-oxidizing bacteria that oxidize nitrite into nitrate nitrogen. N_2O is produced and emitted into the atmosphere during the nitrification process.

At various times, nitrogen flows in various forms through the soil, the Earth's crust, oceans and other bodies of water, plants, animals, and the atmosphere (Fig. 2) Nitrogen is present in the soil in both organic and inorganic forms. It is also present in the human body as organic and inorganic components and in the atmosphere as gaseous components. These cycles flow through both space and time.

2.2 Carbon

Carbon also flows in cycles in various forms, being present in organic form at some times and inorganic form at others, and flows through the soil, the Earth's crust, oceans and other bodies of water, plants, animals, and the atmosphere (Fig. 3). Carbon exists in the ecosystem

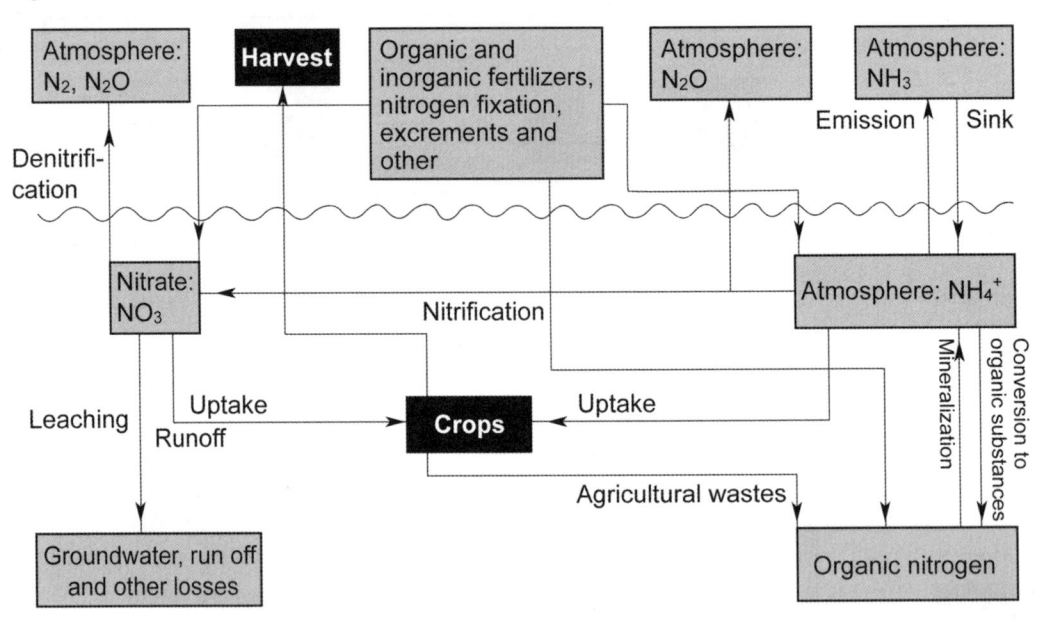

Fig. 3 Nitrogen cycle through the atmosphere, plant, soil and groundwater.

in such forms as carbon dioxide (CO_2), carbon monoxide (CO), methane (CH_4), carbonate (CO_3^{2-}), and hydrogen carbonate (HCO_3^-), and is present in organisms, fossil fuels, sediments, and rocks. The first three gases exist predominantly in the atmosphere, and the two ionic forms are found predominantly in water.

As is the case for nitrogen, the oxidation states of carbon range widely, but between different values: +4 in the form of CO_2 and -4 in the form of CH_4. Carbon also exists as dissolved H_2CO_3, HCO_3^-, and CO_3^{2-} in the ocean and other bodies of water, and as $CaCO_3$, $CaMg(CO_3)_2$, $FeCO_3$, and other mineral forms (in the solid phase) in soils and the Earth's crust.

3. Changes in nitrogen and carbon cycles in the anthroposphere

The birth of the Earth occurred roughly 4.6 billion years ago. After the birth, the various components differentiated into the atmosphere, hydrosphere, crustal sphere, biosphere, and pedosphere. About 10 000 years ago, a new component called the "anthroposphere" appeared as human beings began to significantly affect their environment. Since this time, the anthroposphere has expanded and increased in its impacts to the extent that some now question whether the Earth itself can survive the pressure from its anthroposphere. This question became particularly relevant during the 20th century.

After the birth of the anthroposphere, the nitrogen and carbon cycles were increasingly altered. The first serious changes arose when humans began agricultural production, and the changes accelerated as mankind began to develop affluent materialistic civilizations. Although the nitrogen and carbon cycles have been disrupted by different factors, the following sections present the main causes.

3.1 Nitrogen

The nitrogen cycle has been altered by anthropogenic nitrogen fixation, fertilizer manufacturing, deforestation, reclamation of wetlands, combustion of fossil fuels, and increases in food production. As a result of these changes, various serious phenomena have been initiated: global warming, depletion of the ozone layer, acid rain, soil erosion and changes in nutrient balances in the soil, deterioration of drinking water quality and of groundwater, eutrophication of fresh and salt water, and damage to coral reefs in the ocean.

3.2 Carbon

Carbon cycles have been changed primarily by the combustion of fossil fuels, by deforestation, and by increases in food production. As a result, global warming, acid rain and other atmospheric phenomena, soil erosion, and changes in nutrient balances in the soil have been occurring.

4. Environmental changes caused by changes in the nitrogen and carbon cycles

4.1 Climate change and global warming

Climate change occurs as a result of the interactions of many factors, but currently, it is most obviously being caused by global warming. The current global warming is being caused by human activities and has resulted from increased concentrations of greenhouse gases in the atmosphere. The primary greenhouse gases include carbon dioxide (CO_2), methane (CH_4), nitrous oxide (N_2O), halogenated methane derivatives (e.g., CH_3Br), and CFCs. Carbon dioxide produced by the combustion of fossil fuels is the biggest single contributor to global warming, but the second biggest contributor is methane emitted by paddy fields, livestock, and other sources. The third biggest contributor is nitrous oxide, which is mainly emitted as a consequence of the application of nitrogen fertilizers to arable land.

The effects of global warming have been observed everywhere, but nowhere as dramatically as at the North and South Poles. According to a report by the Worldwatch Institute, the total area of surface ice covering the Arctic Ocean has decreased by about 6% in the 18 years from 1978 to 1996. The thickness of the ice has also decreased dramatically, from an average of 3.1 m around 1970 to an average of only 1.8 m by the mid-1990s. At the South Pole, which holds 91% of all the world's ice, three ice shelves have collapsed, glaciers are diminishing at an alarming rate, and huge icebergs have begun drifting northwards over the past 10 years.

In Japan, too, the average annual temperature continues to rise. In September 1999, the average temperatures in Tokyo and Osaka hit all-time highs of 26.2 and 27.2 °C, respectively, which were similar to temperatures at Amami-Oshima (26.6 °C) and Naha (27.2 °C). Japan has continued to become warmer for a long time. In most areas of Japan, the average annual temperature has risen by more than 2 °C over the past 100 years. The rate of increase has been greatest in Tokyo, where the average has risen by 2.9 °C.

The Japanese Meteorological Agency publishes meteorological information such as average monthly temperature, precipitation, and hours of sunlight at various locations in Japan from Wakkanai to Minami-Daitojima island. In February 2002, the average temperatures from Wakkanai to Hakodate in Hokkaido ranged from 0.1 to −4.3℃. The difference between these average temperatures and those in an average year (an increase) was large, ranging from 1.9 to 4.1℃. The differences (also an increase) at Irkutsk in Siberia, at Yuzhno-Sakhalinsk in Sakhalin, at Vladivostok on the eastern shore of the continent, and at Ulan Bator in Mongolia were 5.4, 3.7, 4.6, and 3.2℃, respectively.

The average temperature in March 2002 hit a record high in various parts of Japan. For example, the average in Tokyo was 12.2℃, 3.3℃ higher than in an average year. The record high temperature in January continued until the spring, and greatly influenced both plants and animals. For example, cherry blossoms appeared earlier than usual at many locations. In the year, the Meteorological Agency reported that the higher-than-usual temperature would continue for a while even in April.

According to the Office of Statistics of the Observations Division of Japan's Meteorological Agency, the average monthly temperature in March 2002 hit a record high at 95 observation points — nearly two-thirds of the 149 observation points that the Agency monitors. The average was higher by more than 3℃ in the Kanto region. The previous record high recorded in Tokyo in 1990 (10.6℃) was beaten in March 2002 by 1.6℃ — the highest temperatures recorded since observations began in 1923. In Osaka, too, the average temperature (11.6℃) was 2.6℃ higher than in an average year, a new record that exceeded the previous record by 1℃. This was the highest temperature recorded since observations began in 1883.

The surface of Lake Suwa, in Nagano Prefecture, freezes each winter, and ice rises and falls along cracks when temperatures drop below a certain level. Since ancient times, local people have believed that this phenomenon is a sign that the god of the Suwa Shrine has arrived and is walking on the surface of the lake. (This is called "the crossing of the god".) Records of this phenomenon have been kept since the Edo Era (1603-1867) at the Yatsurugi Shrine. According to the shrine's records, the absence of a god crossing was observed in only 19 years of this 264-year period (an interval of roughly once every 14 years). In the 44 years of the Meiji Era (1868-1912), a similar periodicity was observed, with no god crossing observed in only 3 years (i.e., once every 14 years). But in more recent times, the frequency of the phenomenon has decreased: In the 13 years of the Taisho Era (1912-1925), only two years had no god crossing (once every 7 years); in the 19 years of the pre-war and wartime Showa Era (1926-1945), only two years had no such event (once every 10 years), and in the 42 years of the post-war Showa Era (1946-1988), no god crossing was observed in 13 years (once every 3 years). In the first 15 years of the Heisei Era (1989-2003), no god crossing was observed in 12 years (once every 1.25 years). This historical evidence clearly illustrates the occurrence and increasing severity of global warming.

As global warming continues to advance, it will become impossible to avoid serious damage to agricultural production as a result of drought and other phenomena at various places. In addition, concern has been raised over the incidence of insect pests, which often become more damaging at higher temperatures. According to a 1998 prediction by the Hadley Center for Climate Prediction and Research in the United Kingdom, if global warming continues, food production will decrease drastically in Africa and North America owing to an increasing occurrence of abnormal weather and other, related factors.

Compounds that contain nitrogen and carbon can increase global warming; these include N_2O, CO_2, and CH_4. These compounds are released by agricultural practices, and livestock, animal wastes, nitrogen fertilizers, flooding of paddy fields, and biomass combustion are particularly serious sources of these compounds.

4.2 Ozone development in the troposphere

Ozone in the troposphere contributes greatly to global warming because, ozone acts as a greenhouse gas. The hydroxide (OH^-) radicals generated by the photolytic reaction between ozone and water vapor control the half-life of other greenhouse gases, including CH_4 and CFCs, in the troposphere, and ozone is also indirectly involved in global warming. An increase in the concentration of ozone in the troposphere directly damages vegetation such as forests and agricultural crops, because ozone is a strongly acidified gas. Because changes in the concentrations of OH^- and HO_2^- radicals also affect the speed of sulfate and nitrate generation, tropospheric ozone is also deeply but indirectly involved with the generation of "acid rain".

On the other hand, ozone in the troposphere helps to prevent harmful ultraviolet rays from reaching the surface of the Earth. These rays are ordinarily intercepted by stratospheric ozone, but levels of stratospheric ozone have decreased dramatically in recent decades. Ozone in the troposphere is thus potentially beneficial in the sense that it eases effects of harmful ultraviolet rays on living organisms and thus on ecosystems. Nonetheless, this benefit is likely to be outweighed by the negative impacts of ozone near the Earth's surface.

Ozone in troposphere is important gas related to such phenomena of the earth in the atmosphere as global warming, acid rain and the depletion of ozone layers in stratosphere.

In recent years, increasing ozone concentrations in the troposphere have been frequently reported, and this change has become a serious issue. The Frontier Observational Research System for Global Change announced recently that ozone in the subtropical westerlies (a broad circulation pattern near the equator) moves between continents and promotes global warming. It is widely known that artificially generated NO_x compounds greatly increase ozone development in the troposphere. Among other agricultural activities, fertilizers, livestock wastes, and biomass combustion are the main sources of NO_x emissions.

4.3 Acid rain

Chemical compounds that contain nitrogen, carbon, and sulfur turn into such organic acids as formate and acetic acid, into metasulfonic acid, and into nitric acid when combined with oxygen in the air. When about 350 ppm carbon dioxide in the atmosphere (the current average concentration) is dissolved in deionized water, the pH level of the water reaches about 5.6 at equilibrium (i.e., mildly acidic). Rain whose pH is less than 5.6 is defined as acid rain.

The pH level of rain water is determined not only by the concentration of acids but also by the levels of basic (alkaline) substances, including ammonia and compounds based on calcium, sodium, and magnesium. About 60% of the anions in rain water in Europe are sulfate ions, and about 30% are nitrate ions. In rain water in Japan, nitrate ions are often present at even higher concentrations. In both cases, the main causes of the acidification of rain water are these sulfates and nitrates. It is thus fair to say that their precursors (sulfur dioxide, other sulfur compounds, and nitrogen oxides) are directly responsible for acid rain. Nitrogen oxides are emitted by the combustion of fossil fuels and biomass as well as by soil organisms. Among agricultural practices, the combustion of biomass, the use of fertilizers, and livestock wastes all cause nitrogen oxide emissions.

4.4 Destruction of stratospheric ozone

Because chlorofluorocarbons (CFCs) are chemically stable, they eventually move into the stratosphere from the troposphere. There, they are broken up by ultraviolet rays, releasing chlorine atoms that combine with the oxygen atoms in ozone and destroy ozone molecules. Sherwood Roland and Mario Molina first identified this phenomenon in 1974. Later, it was found that nitrous oxide (N_2O) emitted from nitrogen fertilizers and methyl bromide (CH_3Br), used as a soil fumigant, also destroy stratospheric ozone.

In 1982, a British research group reported a decrease in the ozone layer over the South Pole. In 1985, an actual ozone "hole" (a zone with low or undetectable levels of stratospheric ozone) was observed over the South Pole. In 1989, a similar ozone hole was detected over the North Pole. The Japanese Meteorological Agency reported in September 2000 that the ozone hole over the South Pole reached its biggest size since observations of this phenomenon began, growing to encompass a region twice the size of Antarctica (29.18 million m^2).

What will be the result of the destruction of the ozone layer and the emergence of widespread ozone holes ? A much greater proportion of the ultraviolet rays in sunlight will reach the Earth's surface. This radiation damages DNA and proteins. Because oceanic phytoplankton are highly sensitive to ultraviolet rays, marine ecology will be adversely affected, and particularly components of the food chain involving the fish and shellfish that feed on these plankton. There are also concerns about the effects of increased ultraviolet radiation on the yield and quality of agricultural crops as well as on the natural global ecosystem.

There are also serious concerns about changes in the Earth's climate due to changes in the heat balance of the atmosphere. This will be discussed later. Furthermore, fundamental problems of the global environment may occur. For example, life has created an environment in which it can sustain itself by generating an ozone layer (a process that required the evolution of photosynthesis and the production of oxygen), and the ecosystem itself may be destroyed by destruction of that ozone layer.

Compounds that contain nitrogen and carbon and that greatly affect the depletion of the ozone layer include N_2O, CFCs, and CH_3Br. In the context of agriculture, these compounds are generated primarily by the use of nitrogen fertilizers, by livestock wastes, and by soil fumigants.

5. Environmental problems that involve nitrogen and carbon

In this section, I go beyond the discussion of the abovementioned environmental problems caused by changes in the nitrogen and carbon cycles and discuss several additional environmental problems that undermine the ecosystem and that involve nitrogen and carbon.

5.1 Invasion by exotic species

"Biological invasions", which take the form of the penetration of non-native organisms into an ecosystem, are probably the least-visible environmental problem and the one that is most difficult to predict. They are also one of the most dangerous consequences of global environmental change.

Many creatures have successfully invaded bodies of water, whether oceanic or inland, by traveling with the current, in the ballast water carried by ships, or in containers borne by those ships. The migration of living things far beyond their original habitat as a result of intentional or accidental flow between habitats is an unavoidable consequence of basic human activities.

These biological invasions have greatly influenced maritime industries, forestry, and horticulture, and their influences are continuing to increase. As the recent occurrences of foot-and-mouth disease have shown, invasions are hardly limited to marine organisms, and we have now entered a new era involving "chaos in the ecosystem". Our era features frequent flows of humans and goods, and almost all creatures can now be brought from their original location to an entirely new location. Who can accurately trace the spread of epidemics in humans, animals, and crops ? It is extremely difficult to predict how future crop production, livestock husbandry, and human health will be affected by such invaders, but modern examples such as the invasion of zebra mussels in the Great Lakes ecosystem of North America suggest that the consequences will not be favorable.

5.2 Dioxins

There is an increasing problem caused by contamination of soils and crops by dioxins?polychlorinated dibenzo-*p*-dioxin, polychlorinated dibenzofurans, and coplanar PCBs. Dioxins are emitted when substances that contain chlorine are burned. Various materials, including chlorine-containing bleached paper, plastics, combustion exhaust, and garbage, contain or emit dioxins. Dioxins emitted from garbage incinerators fall to the ground in the flue gas and pollute soils and crops. The pollution spreads further, from rivers into bodies of fresh water and oceans. Dioxins emitted at one place travel in the air or on oceanic currents and spread across the entire Earth. In the end, they enter the human body through the food chain; therefore, contaminated crops and animals are not suitable as food. Such problems have already occurred in Belgium, in other European countries, and in Japan.

5.3 Genetically modified crops

Genetically modified (GM) crops represent one of the biggest concerns related to agricultural production in the 21st century. The total area planted with such crops around the world in 2000 was estimated at 43 million hectares. This represents an area more than 25 times the area in 1996. The United States accounts for three-quarters of the total planted area of GM crops. More than one-third of the soybean yield in the United States in 1998 was from GM crops. About one-quarter of the corn (maize) and about one-fifth of the cotton was also genetically engineered. In addition to the United States, Argentina and Canada have a high proportion of their crop yields from GM crops. In the same year, more than 50% percent of the soybean yield in Argentina and of the canola (rapeseed) yield in Canada were genetically modified. These three countries account for 99% of the world's planted GM crops. Worldwide, 52% of the area of GM crops is used for soybean production and 30% is used for corn production. The rest is mainly for cotton in the United States and canola in Canada.

The safety of GM crops has been repeatedly questioned. In 1998, the Rowett Research Institute in the United Kingdom reported weakened immune systems in rats that had been fed GM crops over a long period of time. Moreover, transgenic crops containing genes for the production of a natural pesticide (Bt toxin) also have shown an ability to kill creatures other than the targeted pests. Because the elements that function as pesticides may remain in the soil after harvesting, soil pollution is a possibility. If weeds pollinated by insects grow around genetically modified crops, they may also acquire the pesticide gene and become able to kill insects that feed on them, thereby disturbing the ecology of these insects and of animals that rely on these weeds. Crops engineered for herbicide resistance may also cause problems. In Canada, weeds growing around such crops were found to have gained resistance to the herbicides in 1998, only 2 years after the introduction of these crops.

As such cases show, we must closely monitor the production and future use of GM crops that were touted as a "savior of the world against food shortages" when the new crops were first developed.

5.4 Other factors

Nitrogen and carbon are also related to such environmental problems as air pollution, water contamination, desertification, salinization, contamination of marine ecosystems, forest fires, and massive floods, which all have a large influence on food production.

6. Examples of sustainable agricultural activities

6.1 A nitrogen example: N_2O emitted from nitrogen fertilizer

Many strategies for reducing N_2O emissions have been proposed: (1) match the N supply to the crop's demands; (2) "close" the N cycle to prevent leakage into other ecosystems; (3) use advanced fertilizer technologies; and (4) optimize tillage, irrigation, and drainage, in which gaseous emissions related primarily to cropping systems could be minimized. A number of field studies have been conducted with nitrification inhibitors that could potentially decrease N_2O emissions. A few studies have also demonstrated the potential of using controlled-release fertilizers to decrease N_2O emissions.

6.2 A carbon example: CH_4 emitted from rice paddies

Possible strategies for mitigating the potentially large CH_4 emissions from rice paddies are based on controlling the production, oxidation, or transport of this gas. The total emission of CH_4 during the cultivation period can be reduced by 42% to 45% by using short-term drainage rather than leaving paddies continuously flooded in the fallow season. The addition of soil amendments such as nitrate, iron-containing materials, and mineral fertilizers can mitigate CH_4 emissions. The mitigation of CH_4 emissions also requires that the quantities of organic amendments be minimized, at least during the flooded season. Stimulation of composting of organic amendments also appears to be a promising mitigation option.

7. Interactions between adverse global changes and nitrogen and carbon : future research

As discussed earlier, many environmental problems arise from disruptions of the nitrogen and carbon cycles. As a result of these pressures on the global environment, food security and environmental conservation have become serious problems. We now face a situation that humanity has never experienced in the past.

Earlier in this paper, I presented the phenomena responsible for deterioration of the environment individually, as if each factor acted in isolation. However, these phenomena clearly interact. In this section, I discuss possible combinations of the phenomena and the interactions related to agricultural production. I focus on these interactions from the viewpoint of the nitrogen and carbon cycles.

7.1 Global warming + ultraviolet rays + nitrogen and carbon

As the air of the lower atmosphere grows warmer owing to global warming, the temperature of the stratosphere decreases, particularly over the South Pole. The rate of decrease in the ozone layer is accelerated by the cold stratosphere because the rate of reaction between the chlorine in CFCs increases as temperature decreases. The ozone layer over the North Pole will also gradually become thinner as global warming continues to advance.

7.2 Global warming + acid rain + ultraviolet rays + ozone in the troposphere + nitrogen and carbon

In the eastern part of Canada, the volume of water flowing into lakes has been decreasing over the past 20 years owing to a mild drought combined with global warming. Because the volume of organic sediments carried by river waters decreases as the flow of water weakens, the clarity of the water in these lakes increases. This means that a larger proportion of the ultraviolet rays that reach the surface of a lake owing to the depletion of the ozone layer penetrate increasingly deeply into the water as the water clarity increases. For example, a lake in which ultraviolet irradiation formerly penetrated to a depth of 20 to 30 cm may now have a penetration depth of more than 3 m. Simultaneously, acid rain has significantly affected the lake ecosystems of Canada and the northern parts of the Eurasian continent. Destruction of the ozone layer causes a larger volume of sedimentation of dissolved organic matter.

7.3 Global warming + nitrogen

Nitrogen pollution affects the ecology of continental areas, and especially that of forests. Because nitrogen pollution weakens the ability of forests to absorb carbon from the air, it can contribute to a decline in the growth of temperate forests.

7.4 Global warming + decrease in habitat + invasion of exotic species + nitrogen and carbon

Nitrogen pollution, combined with changes in water temperatures as a result of global warming, is believed to be responsible for decreases in the extent of living coral reefs around the world.

7.5 Global warming + infectious diseases + nitrogen and carbon

Even a very small rise in the minimum temperature of a region as a result of global warming can promote the invasion of pests into a new territory. This has immediate and direct consequences for humans, since (for example) the seawater along warm coastal areas can become a breeding ground for the organism responsible for cholera, especially when the water is also contaminated with nitrogen.

7.6 Global warming + forest fires + nitrogen and carbon

Climate change has caused changes in the frequency and severity of forest fires as a result of global warming and related consequences such as more frequent drought. More frequent occurrences of such fires will increase the volume of carbon emitted into the atmosphere, thereby increasing the greenhouse effect. This burning will also cause nitrogen loss in soils and vegetation, plus increased nitrogen emissions into the atmosphere, due to volatilisation and combustion. Climate change and forest fires thus interact, and can multiply each other's effects.

7.7 Global warming + water + nitrogen and carbon

The conversion of dry land into highly productive agricultural land by irrigation permits increased use of nitrogen fertilizers. The consequences of this increased fertilization have already been discussed, but include increased nitrogen emission into the atmosphere and increased nitrogen outflow into bodies of water. Moreover, the use of irrigation has significant effects on soils (e.g., salinization), local heat balances (due to increased evapotranspiration), and water quality (due to leaching of pollutants into the water).

7.8 Deforestation + nitrogen and carbon

Deforestation increases decomposition rates in soil organic matter, plus increases soil erosion and the emission of soil nitrogen and carbon into the atmosphere, with the feedback effects noted earlier in this paper.

7.9 Invasion of exotic species + nitrogen

Nitrogen pollution of grasslands promotes the overgrowth of strong exotic weeds. Nitrogen contamination of forests weakens the resistance of both native and exotic tree species to pests.

Conclusions

The Japanese writer Sawako Ariyoshi wrote Multiple Contamination in 1975, 12 years after the publication of Silent Spring by American writer Rachel Carson. Ariyoshi's story provides many examples of the feared interactions among individual chemicals that will adversely affect the environment. Such fears cannot be downplayed given the changes in the global environment that we are currently observing. As the examples of "global warming" and "ultraviolet rays" show, global warming leads to depletion of the ozone layer.

Sound maintenance of our environment must become the planning goal of the upcoming 100 years and must lead to a revolution in the implementation of our environmental practices. For example, it takes more than 100 years for a nitrate nitrogen concentration of more than 10 ppm in groundwater to decrease to 1 ppm and a similar time period to return carbon dioxide levels (currently as high as 380 ppm in some areas) to a level of 280 ppm, the level that existed at the time of the Industrial Revolution. Without taking dramatic steps, beginning now, we cannot be optimistic about the continuing health of our environment.

However, we seem to have finally begun to understand the necessity for acting decisively to reverse the current environmental crisis. For example, even though it required 40 years to succeed, Carson's Silent Spring triggered many of today's activities to protect our environment. This trend has accelerated. Twelve years have passed since the Intergovernmental Panel on Climate Change (IPCC) formulated the original draft of their first report (Climate Change, The Scientific Assessment) at Harvard University. Since then, the second and third reports have been compiled and international politics can no longer be discussed without discussing environmental problems related to political decisions.

In Japan, too, there is a similar trend. Measures to promote environmentally sustainable agriculture started with the "direction of new food, agriculture and rural policy" in 1992 and the "Basic Law on Food, Agriculture and Rural Areas" in 1999. Later, in response to such developments, three additional environmental laws were enacted (the Sustainable Agriculture, Fertilizer Control, and Livestock Animal Waste Management laws). Private corporations began to compete with each other to acquire ISO 14001 environmental certification. The active involvement of Japan in the Conference of Parties (COP) to the U.N. Framework Convention on Climate Change also reflects the trend towards increased environmental responsibility. Such moves can lead to the discovery of new scientific facts that are required for the understanding of the problems and to the further development of technologies and systems that can resolve these problems. These contributions will depend heavily on constant, sincere efforts by researchers and on protecting the continuity of research.

The global environment has no future without such efforts. In brief, sustainable agricultural practices that conserve the environment in all countries will lead to conservation of the global environment. Only by adopting these practices can the "constant" nature that Shiba wrote about persist into our future.

References

Brown, L. (1996 to 2001) Earth white paper 1996 to 2001. (ed. By Y. Hamanaka). Diamond Inc. (in Japanese)

Brown, L. (2000) Lester Brown's environmental revolution: for environmental policy in 21st century. (ed. by H. Matsuno). Sakuhokusha. (in Japanese)

Jackson, A.R.W. and Jackson, J.M. (1996) Environmental Sciences. Longman.

Minami, K. (2002) World's environmental resources and problems that affect food security. *Agriculture and Horticulture* 77: 5−9 (in Japanese)

Minami, K. (2002) Nitrous oxide emission: Sources, Sinks and Strategies. Encyclopedia of Soil Science, (ed. by R. Lai), Marcel Dekker: 864−867

NIAES (2002) Information: Agriculture and environment No. 25 etc., National Institute for Agro-Environmental Sciences, http://www.niaes.affrc.go.jp. (in Japanese)

Sankei Shimbun (ed.) (2000) What is important in life. Toyo Keizai Inc. (in Japanese)

Yagi, K. (2002) Methane emission in rice, mitigation option for. Encyclopedia of Soil Science, (ed. by R. Lai): 814−818

Environmental and ecological aspects of Korean agriculture

Ki-Cheol Eom[a], Tae-Wan Kim[b] and Jae E. Yang[c]

[a] National Institute of Agricultural Science and Technology, Suwon, Korea
[b] College of Agriculture and Life Sciences, Hankyong National University, Ansung, Korea
[c] College of Agriculture and Life Sciences, Kangwon National University, Chunchon, Korea

Abstract

Korean irrigation methods are selected on the basis of topography, soil texture, water resources, and cost. The main methods involve furrow irrigation, and the use of water-diversion hoses, mini-sprinklers, and drip-watering. On slopes, serious drought stress affects crops owing to high runoff and low water infiltration into the soil, and soil erosion increases dramatically, leading to low soil fertility, a gradual reduction of effective soil depth, and ultimately sediment deposition in river bottoms. A guidebook has been published for optimum water management based on soil characteristics and average rainfall in most agricultural regions. This book attempts to teach Korean farmers optimum fertilizing and irrigation strategies. The Korean government is now implementing an agricultural management plan to achieve soil conservation through the use of methods such as contour cropping, deep tillage, the creation of grass zones and catch canals, terrace cultivation, and gravel belts. This paper reports the results of our research related to water and soil conservation and summarizes the current situation in Korea.

Keywords : irrigation, fertilization, soil conservation, sustainable management

1. Background

The structure of Korean agriculture during the last 40 years has changed from subsistence farming to environmentally sound sustainable agriculture (Fig. 1). The general goals of agriculture in recent historical periods were as follows: small-scale subsistence farming of arable land (until the 1960s); enhancement of crop productivity (1960s–1970s); profit-oriented farming (1970s–1980s); multiple-phase agriculture, including the use of cash crops, horticulture, and livestock production (1980s–1990s); relatively large-scale commercial or professional agriculture for broader markets (1990s–2000); and environmentally sound, sustainable agriculture (since 2000).

In the past, more than half of the Korean population lived in rural areas and provided

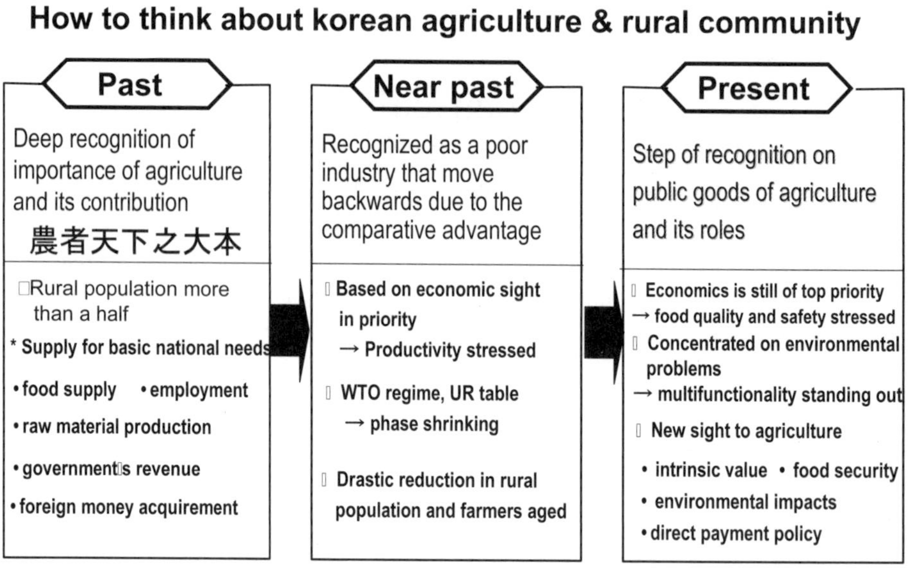

How to think about korean agriculture & rural community

Past	Near past	Present
Deep recognition of importance of agriculture and its contribution 農者天下之大本	Recognized as a poor industry that move backwards due to the comparative advantage	Step of recognition on public goods of agriculture and its roles
▫ Rural population more than a half * Supply for basic national needs • food supply • employment • raw material production • government's revenue • foreign money acquirement	▫ Based on economic sight in priority → Productivity stressed ▫ WTO regime, UR table → phase shrinking ▫ Drastic reduction in rural population and farmers aged	▫ Economics is still of top priority → food quality and safety stressed ▫ Concentrated on environmental problems → multifunctionality standing out ▫ New sight to agriculture • intrinsic value • food security • environmental impacts • direct payment policy

Fig. 1 Chronological changes of perception on agricultural community in Korea

materials to meet their own basic needs and national needs for products such as food, fiber, and raw materials, some of which represented major sources of government revenue and foreign trade. As a result of economic and industrial developments over the past several decades, however, Korean society has begun to consider such rural functioning as a poor industry that operates at a disadvantage compared to other industries. The uncertain economic viability of farming, along with international pressures to open Korea's agricultural markets, has resulted in drastic reductions in rural populations.

At present, a major agricultural policy related to crop production and environmental conservation involves establishing a system of environmentally sound sustainable agriculture and assessing the multiple functions performed by agriculture. The National Institute of Agricultural Science and Technology (NIAST) has played a key role in assessing the multiple environmental functions of Korean agriculture by forming multidisciplinary research teams that have investigated water use, soil conservation and management, fertilizer management, recycling of organic resources, and monitoring of atmospheric changes. This paper describes the results of many of these investigations.

2. Current Status of the Agricultural Environment in Korea

Most Korean paddy fields lie in flat valley bottoms (85.2 %), although some rice is cultivated on slopes (21.4%). The soils of Korean agricultural fields derive mainly from acidic rocks (e.g., granite, granitic gneiss), leading to leaching during intense rainfall. Korea has begun soil conservation using methods that are practical for farmers working on sloping land. The Rural Development Administration (RDA) has developed a financial support plan

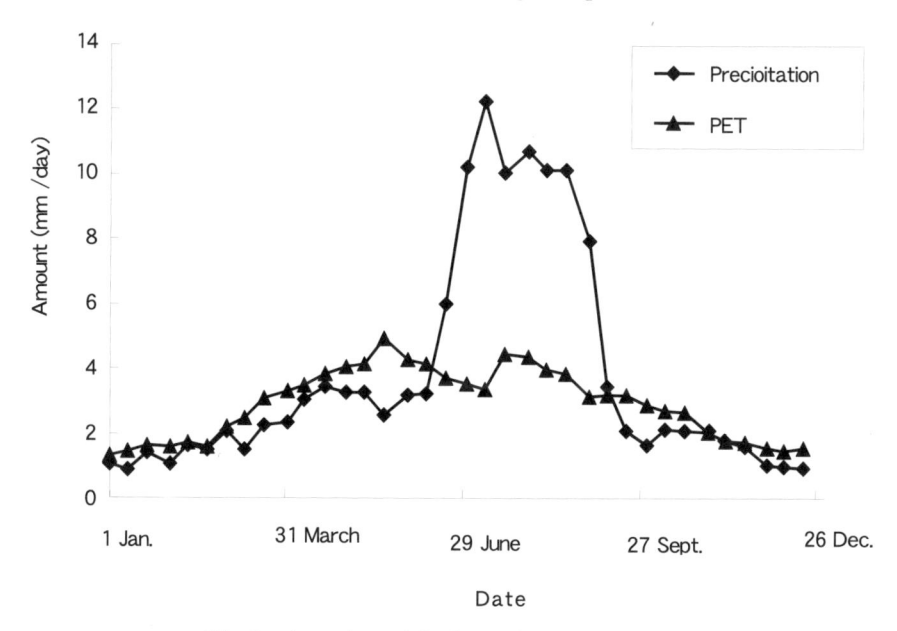

Fig. 2 Annual precipitation and potential in Korea

Table 1 Water resources and availability in Korea compared with the world average

Item	Korea	World average	Ratio
Annual precipitation (mm)	1,274	973	130.9%
Precipitation per capita (ton/person)	2,900	26,800	10.8%

※ Korea is a water-deficient country

on the basis of the environmental factors that affect both small farms and large regional farms.

2.1. Water Use

Korea is considered to be a water-deficient country (Table 1). The biggest problem in managing water use in Korean agriculture is that the majority of the precipitation is concentrated during the summer monsoon season (Fig. 2). The estimated total water resource is 126.7 billion ton, but the total water use is only about 30.1 billion ton (24 %), of which 63 % is available for agriculture. Figure 3 describes the water resources and balances in a typical Korean paddy field. Half of the irrigation water or precipitation is estimated to infiltrate the soil; the rest mainly undergoes evapotranspiration. During the rice cultivation period, the total water supply is equivalent to 2177 mm (1266 mm of irrigation and 947 mm of precipitation), whereas total water consumption is composed of evapotranspiration (863 mm), runoff (224 mm), and infiltration (1041 mm), with a mean flooded depth of 45 mm.

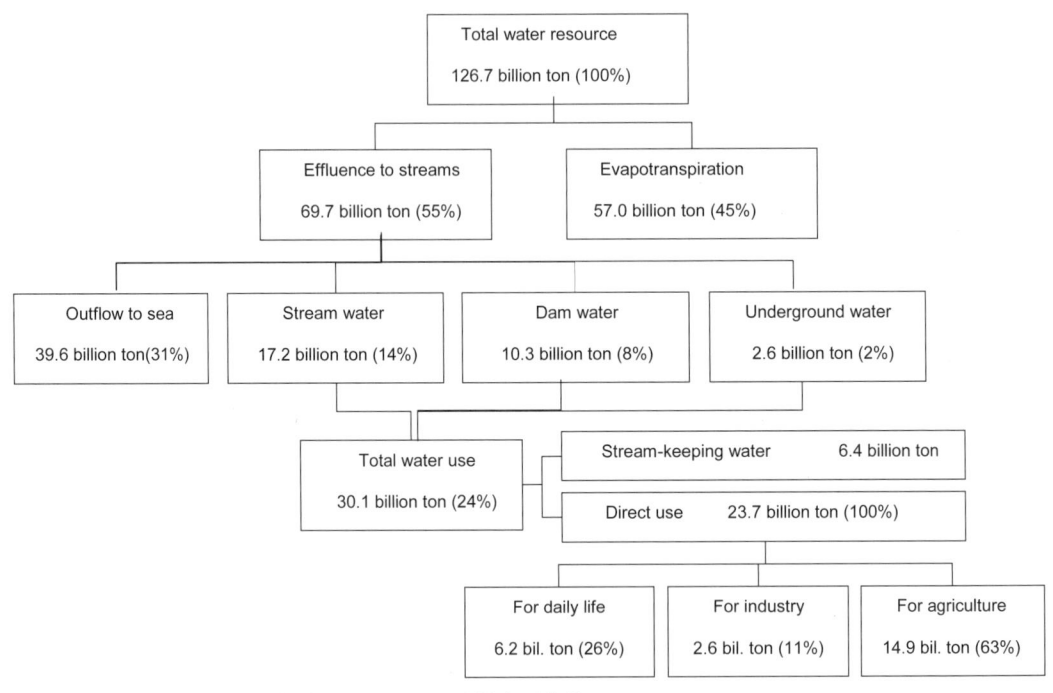

```
                    Total water resource
                    126.7 billion ton (100%)
                    ┌──────────┴──────────┐
              Effluence to streams    Evapotranspiration
              69.7 billion ton (55%)  57.0 billion ton (45%)
```

Fig. 3 Water resource and water use balance in Korea

(diagram continues with: Outflow to sea 39.6 billion ton (31%), Stream water 17.2 billion ton (14%), Dam water 10.3 billion ton (8%), Underground water 2.6 billion ton (2%); Total water use 30.1 billion ton (24%); Stream-keeping water 6.4 billion ton; Direct use 23.7 billion ton (100%); For daily life 6.2 bil. ton (26%), For industry 2.6 bil. ton (11%), For agriculture 14.9 bil. ton (63%))

; 10 % of agricultural use = 57 % of industrial use use, 24% for daily life

2.2 Soil Loss and Management Strategy

More than 60 % of Korean lands are forested, and only 21 % are cultivated paddies and uplands. Uplands occupy about 7 % of the total land, and about 62 % of the uplands are on slopes greater than 7 %. Soils on sloping uplands are subject to intensive land use, with a high

Table 2 The selected management options currently practiced to control soil erosion

Practices	Description
Contour faming	Construction of furrows of 10~15cm in height (sowing or transplanting on the furrows)
Deep plowing+Contour farming	Deep plowing at 20cm over in depth in case of soils with higher content of clay or with thin layer of top soil Construction of furrows in the direction of contour
Green strip	Construction of green strips of 1.0~1.2m in width (Length of slope at interval of 20~30m)
Drainage canal	Construction of drainage canals using gravel or grasses in the direction of contour (Length of slope at interval of 20~30m)
Terraced land	Construction of terraces using gravel or grasses in the direction of contour (Length of slope at interval of 20~30m, Height of terraces 30~50cm)

input of agrochemicals, and are vulnerable to soil erosion. Development of an environmentally sound land-management strategy is essential in order to permit a sustainable production system in these uplands.

In sloping uplands, soil erosion and thus loss increase dramatically, resulting in low soil fertility, a reduction in the effective depth of the surface soil, and a rise in the level of river bottoms due to sediment deposition. These changes have detrimental effects on water quality within the agricultural watershed. Several management options for soil conservation have

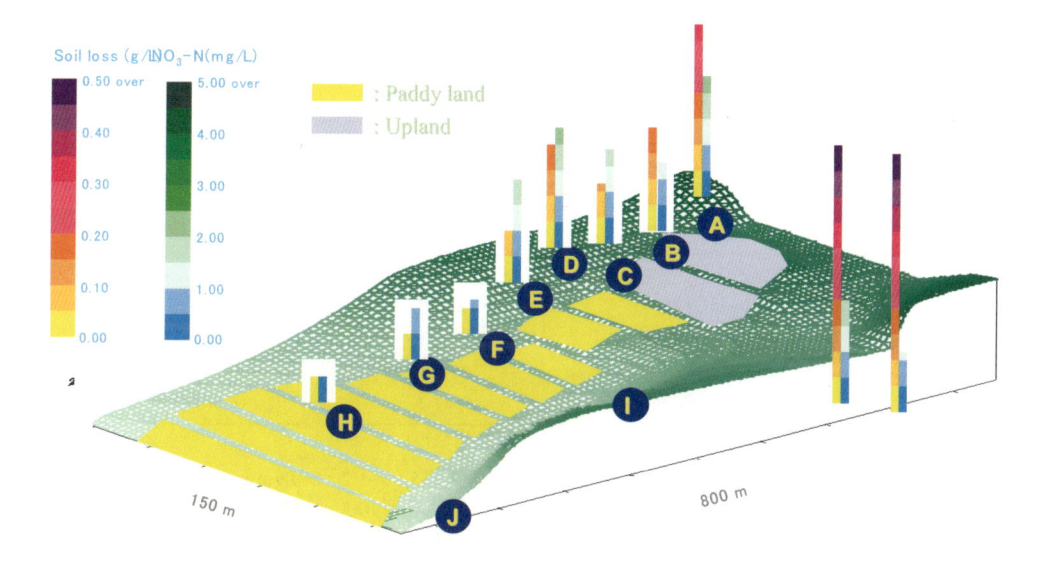

Fig. 4 Soil and nitrate-N losses one day after heavy rain in an agricultural land

Slope	Area (thou.ha)	Soil loss (thou.ton)	Yearly soil loss (ton/ha/yr)	Grade	Recommanded practices
A (0 2%)	1.5	1.2	0.8	Very good	
B (2 7%)	0.5	2.2	4.2	Good	Contour farming Mulching, Deep plowing
C (7 15%)	2.4	23.6	9.7	Moderate	Green strip
D (15 30%)	3.8	78.5	20.4	Severe	Drainage canal
E (30 60%)	5.4	355.0	65.4	Very severe	Terraced land
F (60 100%)	1.6	127.9	80.5	Very severe	Terraced land
Average	15.3	588.4	38.4	Severe	

Fig. 5 Soil erosion map in Pyeongchang County with various land use types

Table 3 Nutrient losses as a result of runoff, leaching, and soil erosion and the effects of soil textures

(kg/ha)

Item \ Texture	Sandy loam	Loam	Clayey loam
NO_3-N			
Runoff	20.9	23.7	25.8
Leaching	9.0	7.8	7.2
In eroded soil	0.11	0.14	0.17
Sum	30.0	31.6	33.2
P_2O_5			
Runoff	1.42	1.74	2.05
Leaching	0.35	0.43	0.46
In eroded soil	1.83	2.26	3.15
Sum	3.60	4.43	5.66
K_2O			
Runoff	18.8	21.3	20.3
Leaching	7.8	7.1	7.4
In eroded soil	2.3	3.1	4.3
Sum	28.9	31.5	32.0

★ Fertilization N. P_2O_5. K_2O = 290-200-230 kg/ha
★ Concentration in runoff water $NO_3-N : P_2O_5 : K_2O$ = 1~55 : 0~1.2 : 1~38 mg / 100g
★ Concentration in leaches = 1~34 : 0~0.5 : 2~12 mg / 100g
★ Content in eroded soil = 1~5 : 75~212 : 50~141 mg / 100g
★ Before experiment = 20~27 : 90~142 : 124~195 mg / 100g

been practiced on sloping uplands, such as the use of contour cropping, grass or gravel buffer strips, detention weirs or ponds, terrace cultivation, crop rotation, inter-cropping, cover crops during the winter, grassed borders of waterways, diversion drains, and mulching farming (Table 2). As shown in Figure 4, the results of field monitoring show that soil loss in the terraced land after heavy rain differed according to slope, land use type, and crop cultivation. The amount of soil loss ranged from 0.074 to 0.718 g/L. Similarly, leaching of nitrate-N varied as a result of differences in these three factors, ranging from 0.41 to 8.91 mg/L.

Soil texture is closely related to the magnitude of the observed nutrient loss (Fig. 5 and Table 3), which may lead to eutrophication of the watershed. Nutrient loss was higher in clay loams than in other textural classes owing to increased runoff and erosion (Table 3). As shown in the soil erosion map (Fig. 5), soil loss increased dramatically as slope increased. In paddy fields with slopes below 2 %, soil loss was about 10 % of the values in mountainous land.

2.3. Fertilizers

Farmers tend to apply more fertilizers (N, P, and K) than the recommended level (Table 4).

Table 4 Comparison of actual fertilizer consumption with the recommended amounts (kg/ha)

	N	P_2O_5	K_2O	Sum
Total requirements (A)	288	138	182	648
Real consumption (B)	428	173	201	801
Difference (B−A)	140	35	19	193
(ratio : B/A)	(1.49)	(1.25)	(1.10)	(1.24)

* There is still a tendency of heavy fertilization for high yield.
  − Nutrient imbalance, Frequent occurrence of pests, susceptible to natural injuries
* The same level of fertilization as a recommandation is still applied for composite fertilizers nationwidely.

They apply about 1.5, 1.3, and 1.1 times the recommended quantity of these fertilizers, respectively, during crop production. This increases both the economic burden on the farmers and the amount of pollution in agricultural runoff. Recently, in an attempt to avoid these and other problems inherent in the use of chemical fertilizers, many farmers have shifted from conventional farming systems to organic farming. Organic farming is also called "natural" or "biodynamic" farming, and offers advantages in terms of food safety and resource recycling. First, this approach increases the consumption of organically and domestically grown agricultural products and satisfies consumer demand. Second, a variety of organic resources, such as agricultural byproducts, animal wastes, and other organic wastes, can be recycled by organic farming. However, several problems have arisen in terms of productivity, labor and energy inputs, pest and weed control, the quality of organic composts, and environmental pollution.

We have assessed the relationship between optimum fertilization rate and water quality

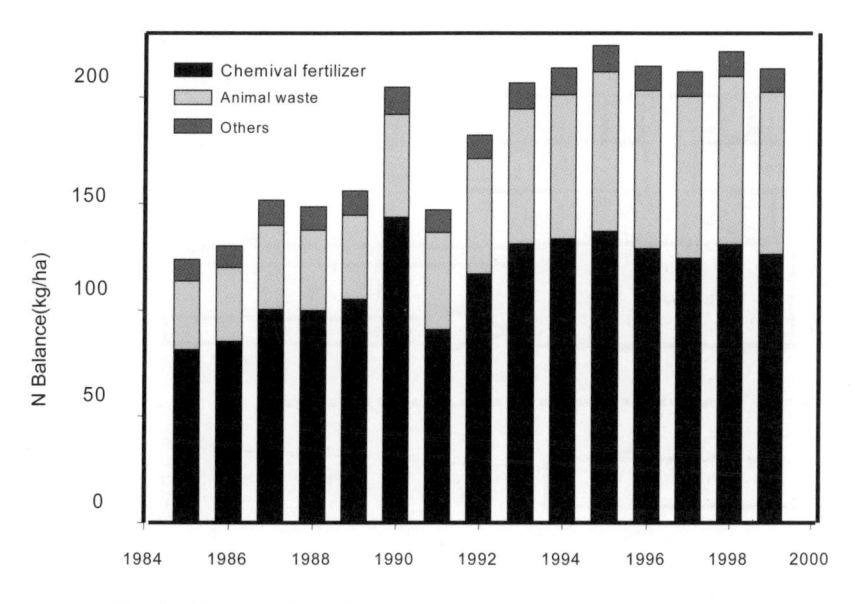

Fig. 6 Changes of N balance among different N sources in Korea

with the goal of minimizing the burden upon the agricultural environment. We estimated crop yields using the following function:

$$Y = A_0 + A(1 + e^{cx}) \quad \text{(Eq. 1)}$$

where Y is water quality; c, A_0 and A are regression coefficients; and x is fertilization rate. Using this function according to the law of diminishing returns, we found that the amount of fertilizer used could be reduced by about 32 %. Furthermore, we observed that crop yield was closely related to the effectiveness of the water management.

2.4. Animal Wastes

In Korea, the amount of animal waste applied to agricultural land has increased continuously (Fig. 6). In 2000, 217 000 tons of nitrogen was input to agricultural land through the application of the animal wastes, compared with 423 000 tons of chemical fertilizers. Total nitrogen removals from soils through crop production and leaching reached 238 000 and 406 000 tons, respectively (Fig. 7).

2.5. Organic Resources

The total amount of nutrients (summation of N, P_2O_5 and K_2O) produced from organic resources was estimated at 1 487 000 tons (Table 5), a value that exceeded the real amounts of nutrients supplied through fertilizers (648 000 tons). Most were derived from animal wastes (823 000 ton) and industrial organic waste (620 000 ton). A proper management strategy must be established for these organic resources from the viewpoint of nutrient management and minimizing the environmental load created by fertilization. Moreover, there are potential risks from relying on organic fertilizers. For example, Table 6 demonstrates that

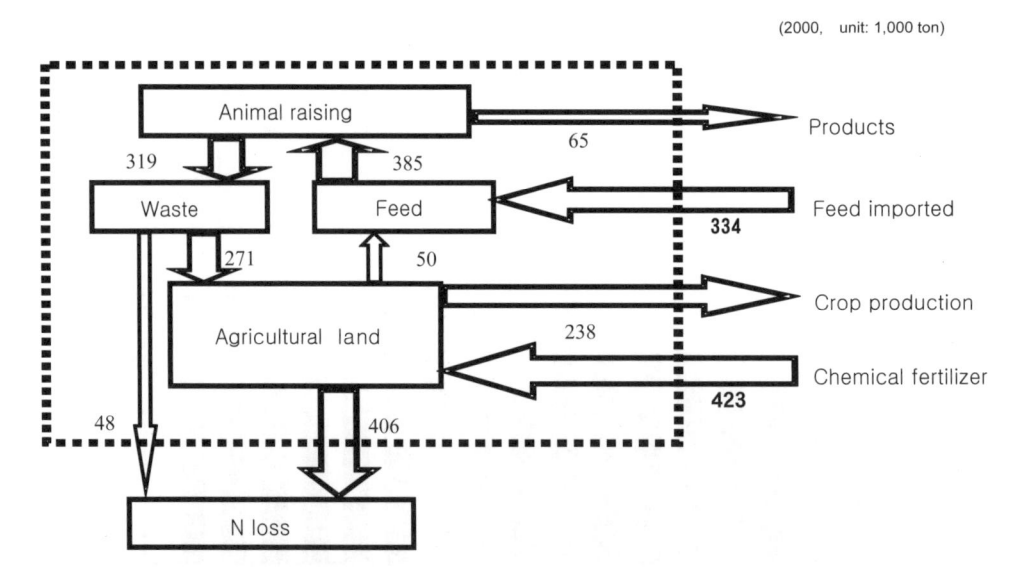

* Amount of imported feed calculated based on domstic feed production

Fig. 7 Nitrogen balance in the agricultural ecosystem in Korea

Table 5 The amounts of nutrient sources produced from selected organic resources in Korea (unit: ingredient, thousand ton)

Species of wastes	N	P_2O_5	K_2O	Total
Animal waste ('98)	310	271	242	823
Food waste ('97)	29	10	5	44
Industrial organic waste ('95)	365	174	81	620
Sum	704	455	328	1,487
Capacity of disposal* ('97)	288	138	182	648

*Exceeding total amount of wastes possibly comsummed in land when estimated based on recommanded rate of N.

Table 6 Fertilization rates from various sources and their impacts on soil characteristics and crop yield

	Application rate (ton/ha)	pH (1:5)	EC (Ds/m)	OM (g/kg)	T-N (g/kg)	AV. P_2O_5 (mg/kg)	Cu (mg/kg)	Zn (mg/kg)	Cr (mg/kg)	Yield index (Raddish)
Soil before		4.9	0.24	10	0.7	17	0.3	0.2	ND	—
Fertilizer		5.4	0.27	15	0.8	137	2	4	0.3	100
Sewage sludge	25	5.8	0.26	21	1.0	298	5	26	0.1	67
	50	5.2	0.56	24	1.3	606	7	43	0.2	66
	100	4.9	0.64	31	1.6	832	10	55	1.9	57
Industry sludge	25	5.4	0.71	22	1.3	125	170	87	7.3	4
	50	4.7	0.83	27	1.8	186	303	71	14.9	1
	100	4.6	0.97	38	2.9	237	269	77	27.5	0
Leather sludge	25	6.3	0.46	28	1.7	22	4	15	16.7	41
	50	6.4	0.78	36	2.8	25	3	26	31.6	51
	100	6.5	2.14	51	4.6	35	3	49	83.4	65
Alcoholic sludge	25	5.1	0.37	30	1.8	152	6	9	0.0	63
	50	4.2	0.89	40	3.1	263	11	10	0.0	81
	100	3.8	1.32	60	5.0	373	15	11	0.0	73
Pig manure compost	25	6.4	0.30	21	1.1	441	9	21	0.1	57
	50	6.4	0.43	30	1.4	902	13	39	0.1	93
	100	6.6	0.77	43	2.3	1,749	15	90	0.2	127

successive applications of organic resources could degrade the soil quality as a result of the accumulation of heavy metals and salts.

2.6. Changes in the Atmospheric Environment

Recently, the world has begun to face an abnormal climate resulting from changes assumed to be caused by the emission of greenhouse gases, SO_x and NO_x deposition, ozone destruction or generation, UV-B radiation, and global warming. In the Korean Peninsula, these atmospheric changes are exacerbated by sand and dust carried by the wind (the so-called "yellow dust") that annually reaches Korea from Asia. The diameter of these dust particles is as small as 11.8 μm. Recent research has shown that the contents of microbes such as bacteria, fungi, and *Bacillus* sp. in the dust were 14, 35, and 21 times, respectively, the levels observed in dust particles under normal conditions.

Table 7 Physiological responses of rice to ozone

	Amount of photosysthesis (mol m^{-2}s^{-1})	Conductivity of stomata (mol m^{-2}s^{-1})	Amount of evaporation (mmol m^{-2}s^{-1})	Soluble carbohydrate (%)	Starch (%)
No treatment	3.2	0.06	1.7	13.1	5.8
O$_3$ (150ppb)	1.3	0.07	1.0	8.7	7.3

O$_3$ (150ppb) ⇒ Decrease rate of photosynthesis, rate of ET, and soluble carbohydrate.

Table 8 Environmental parameters considered in designing an integrated management system for sustainable crop production in Korea

Environmental condition	Characteristics	Consideration
Climate	- Summer: high temp. heavy rain - Winter: low temp. dry season	- Limit cultivation periods and regions
Topology	- Paddy: plain & valley (85%) "below slope 7% (77%)" - Upland: valley & mountainous slope (55%) hill land & mountains (21%) "over slope 7% (66%)"	- Applicable Tech, by various conditions
Soil	- Classification: 378 soil series (paddy 152, upland 125, forest 101) - Physical: Rock (granite, granite gneiss), slope distributions - Fertility, OM, Phosphorus, Potassium, EC: high - Ca, Mg, avail. Silica: low	- Retention of water and nutrient Difference of material flux - Soil acidity - Unbalance of fertility at paddy, upland, orchard, vinylcultivation
Water resources	- Water deficiency: all periods except late June ~ middle Sept.	- Inevitable irrigation
Crops	- Cereals, field vegetables → fruit tree, vinylhorticulture	- Load of environmental pollution due to focusing income
Cropping system	- Vegetable at paddy (22%), load mixed putting first income crops	- Pesticide, fertilizer, labor, management cost
Application tech.	- More applied than standard (1.24 fold)	- Requirement of reforming standard for application rate

The concentration of ozone in the air of agricultural lands near industrial districts has been high since the end of the past century. To cope with this phenomenon, research trials by NIAST have selected ozone-tolerant rice cultivars comparable to the Japanese 'Sasanishiki' cultivar. Some physiological parameters have been considered as selection criteria (Table 7). Deliberating on "Green Round", NIAST has developed an integrated management system for sustainable crop production and environmental conservation by considering several parameters such as topography, soil type, water resources, cropping system, and application technology (Table 8).

3. Perspectives for Future Development

3.1. Introduction of Precision Farming Technology

The new paradigm for environmentally sound sustainable development of agriculture in

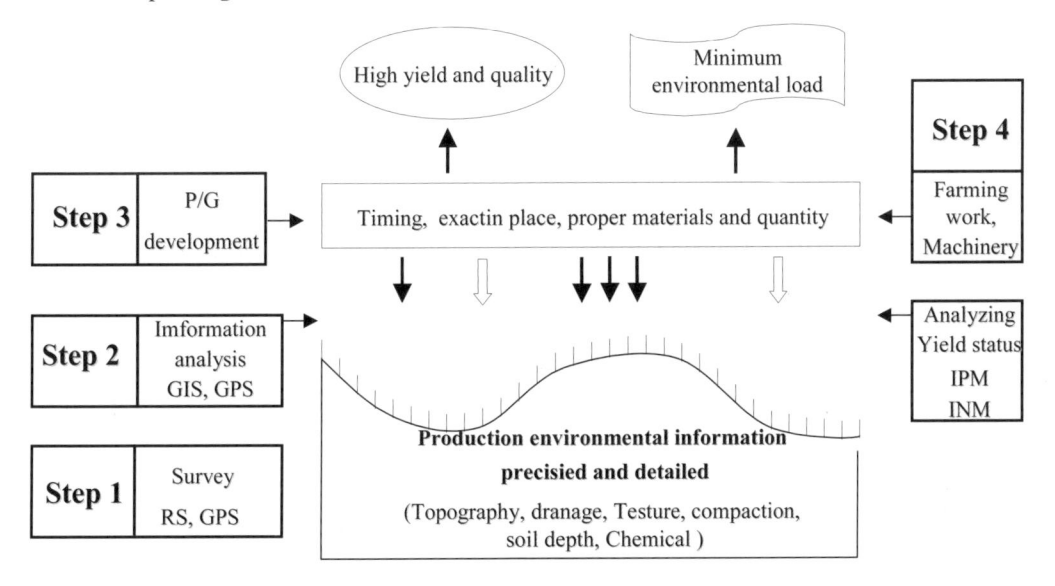

Fig. 8 Schematic flow of the precision farming system for minimum environmental load and high quality production

Table 9 Soil management options for sloping lands that can help to prevent soil erosion

Soil management	Property of subjected area	Area (thou. ha)
Contour culture	Slope 2~7%	190.0
Deep plowing+ contour culture	Clayey clay	122.6
Grass band	Slope 7~15%	124.6
Catch canal	Slope 5~30%	54.4
Gravel band	Alphine	46.6
Terraced land	Over slope 30%	24.0
Total		562.2

Korea is to adopt "precision farming" technology. The goal of the initial stages of adopting precision farming is to construct an information system to collect and interpret data on agricultural parameters such as topography, drainage, soil depth, soil compaction, input chemicals, and other factors that interact with the burden on the environment and the product quality (Fig. 8).

3.2. Prevention of Soil Erosion

The Korean Ministry of Agriculture and Forestry is implementing a subsidiary plan for the prevention of soil erosion on sloping lands (Table 9). This plan is part of a soil conservation strategy addressing soil management practices such as the use of contour culture, deep plowing, buffer strips, detention ponds, terraced land, and gravel bands.

3.3. Irrigation

Irrigation is considered to be one of the crucial factors in sustainable production systems. Figure 9 clearly demonstrates that proper irrigation management combined with fertilization increased crop yields, and that irrigation contributed more to yield increases than did fertilization rates. NIAST has published a guidebook for optimal water management based on soil characteristics, fertilization regime, and average rainfall. In the guidebook, the irrigation demand can be easily calculated using a Microsoft Excel spreadsheet. At present, the irrigation demands of total 38 crops have been established.

3.4. Improvement of Irrigation Water Quality

Various water sources have been used for irrigation, but each has different water quality

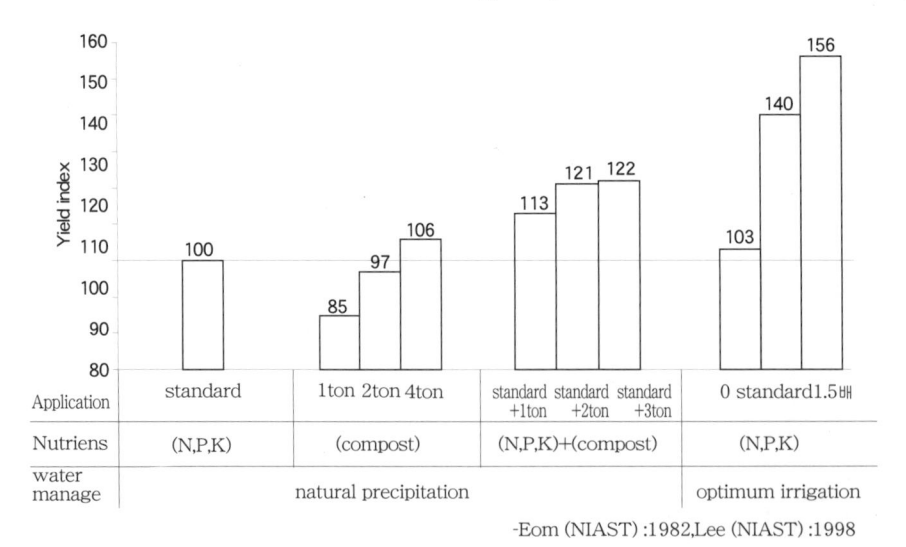

Fig. 9 Comparison of cabbage yields as affected by combination of irrigation and fertilizer managements

Table 10 A proposed water quality standard for agricultural irrigation in Korea

(Unit : mg/L)

	Standard according to elements							
	COD	BOD	SS	DO	T-N	NO_3-N	T-P	Cl
Stream water (IV)	–	below 8	below 100	over 2	–	–	–	
Reservoir (IV)	below 8	–	below 15	over 2	1.0	–	0.1	
Underground water	below 8	–	–	–	–	20	–	250

※ Others: Cd (0.01), As (0.05) below, CN, Hg, organic phosphorus hasn't to detect, Pb (0.1) below, Cr+6 (0.05) below

※ Exception: Using agricultural water, if T-N/T-P ratio is less than 7, don't apply T-P, and more than 16, don't apply T-N.

⟨Amount of side runoff water at vinylcultivation⟩

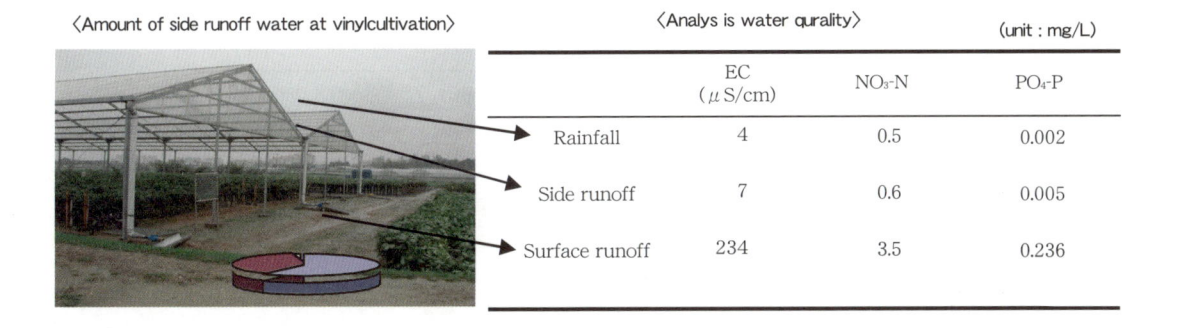

⟨Analysis water qurality⟩ (unit : mg/L)

	EC (μS/cm)	NO_3-N	PO_4-P
Rainfall	4	0.5	0.002
Side runoff	7	0.6	0.005
Surface runoff	234	3.5	0.236

Fig. 10 Construction of storage facility for rainfall and runoff collection, and the analytical data for each water resource

standard, which makes it difficult to manage irrigation resources. On the basis of long-term monitoring data, we developed a revised water quality standard for agricultural irrigation (Table 10). When using reservoir and underground water, COD should be below 8 mg/L and stream water should have BOD lower than 8 mg/L.

3.5. Water Resource Recycling

An NIAST research project assessed the concentrations of NO_3-N and PO_4-P in the runoff from the vinyl cultivation facilities in an effort to determine the potential for recycling water (Fig. 10). The amount of surface runoff and nutrients showed considerable potential for recycling. Another option for enhancing water-recycling efficiency would be to construct an underground reservoir (Fig. 11). The proper dimensions for such a reservoir would be a valley width between 300 and 600 m, a deposit layer thickness of 10 to 20 m, and a slight slope with unweathered bedrock.

3.6. Fertilizer Recommendations

We assessed a new model for optimizing fertilizer application by considering water quality

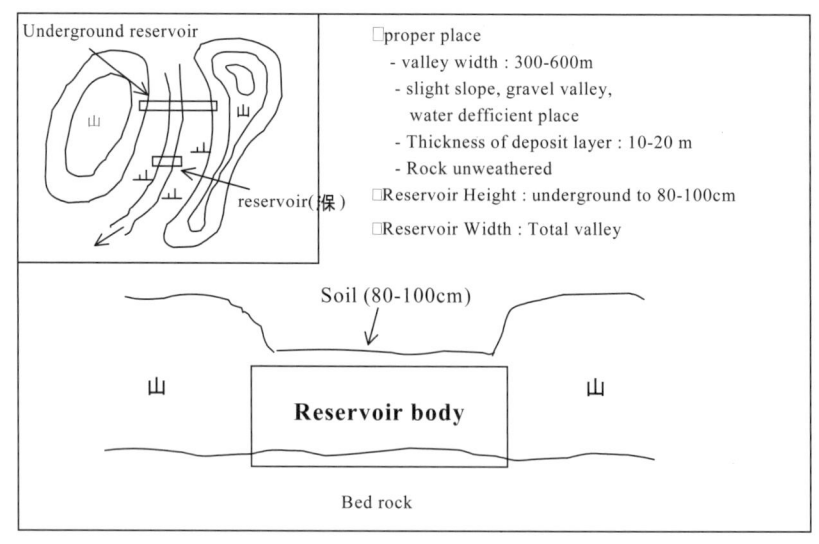

Fig. 11 Schematic model for underground water reservoir in the valley

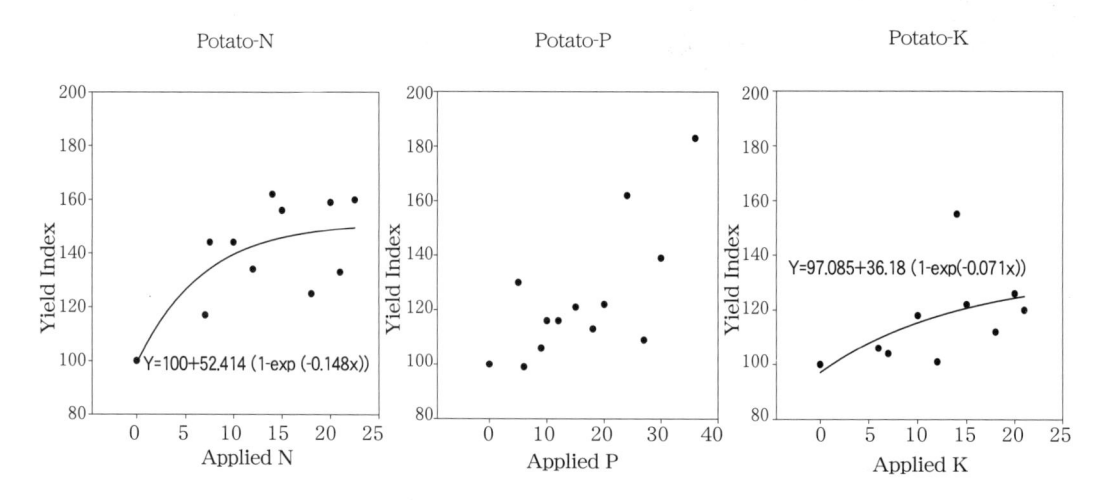

Fig. 12 Assessment of potato yield response to fertilizer rate using different model

so as to minimize the eutrophic burden on the agricultural environment. Crop yields were estimated using two modified functions:

$$Y = A_0 + A(1+e^{cx}) \quad \text{(Eq. 1)}$$
$$Y = A \ln(x) + A_0 \quad \text{(Eq. 2)}$$

where Y is crop yields ; c, A_0 and A are regression coefficients ; and x is fertilizarion rate. Using these functions and the law of diminishing returns, we estimated that the amounts of fertilizer used could be reduced by about 32% for potatoes (Fig. 12). We are currently

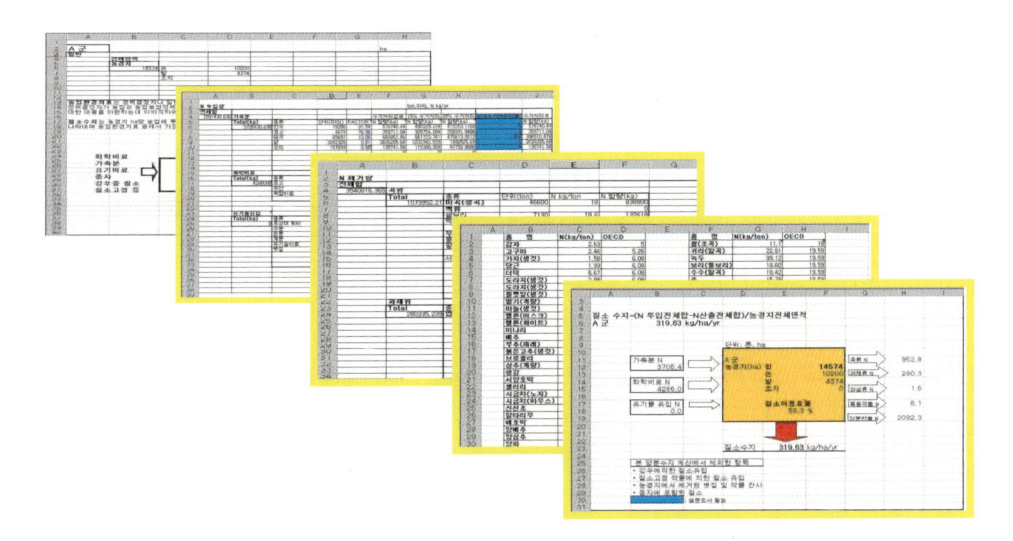

Fig. 13 Excel program for calculation of nutrient balance on a local scale

developing a Microsoft Excel spreadsheet to facilitate the calculation of local nutrient balances (Fig. 13).

3.7. "Fertigation" (Fertilization + Irrigation)

The term "fertigation" has been defined as simultaneous fertilization and irrigation. As a tool for precision agriculture, "fertigation" offers several advantages: lower inputs of water and nutrients, proper management for yield, convenience, reduced labor requirements, increased application efficiency, a more cost-effective approach, decreased nutrient loss, and prevention of soil erosion. A detailed guideline for "fertigation" is under development by NIAST and will be incorporated within the precision agriculture system.

3.8. Recycling Organic Wastes

As a part of national project for environmental conservation, a biogas plant was developed to recycle various organic waste resources (Fig. 14). This specific biogas plant has been designed to recycle manure slurries, and can generate about 600 kWh per day of electricity. This can ultimately reduce electrical costs by ca. 21 600 thousand won per year.

3.9. Energy Balance

Simultaneously sustaining agricultural productivity and conserving the surrounding environment are an important theme in modern agriculture. It is therefore necessary to develop technologies that use all resources more efficiently in order to reduce the need for fertilizers and other agrochemicals that consume large amounts of fossil fuels. Various organic materials have been used as alternatives to chemical fertilizers, but the use of chemical fertilizers or pesticides versus organic materials has not been examined sufficiently in terms of energy balances and material cycling. Among the selected crops we have monitored, the

Table 11 Energy balance for selected upland crops (unit: 10^3 Mcal/ha)

Crop species	Input energy (A)	Output energy			Energy balance (D−A)	Net energy (B−A)	Energy use efficiency (%) (B/A)
		product (B)	residues (C)	Sum (D)			
Soybean	4.1	5.3	6.8	12.1	8.0	1.2	131
Potato	8.7	18.6	7.8	26.4	17.7	9.9	215
Cabbage	10.7	26.0	–	26.0	15.3	15.3	244
Red pepper	13.1	8.7	23.3	32.0	18.9	▽4.4	67

※ Development of low input farming technologies based on solar energy flux and energy balance

Fig. 14 Development of the biogas production system

input energy for red pepper production is highest and the energy-use efficiency is lowest (Table 11). Energy balances can be influenced by many environmental factors, which should be incorporated into models for assessing the economics of sustainable agricultural management.

4. Conclusions

In this paper, we have summarized the current status of Korean agriculture in terms of its environmental and ecological aspects. Agriculture has multiple functions, both as a food supply and as an environmental industry. Farmers are thus important guardians of the land and important actors in environmental conservation. To establish environmentally sound and sustainable agriculture, precision farming should be adopted. In a new agricultural paradigm for the 21st century, research on material circulation between the soil, water, crops, and atmosphere should be conducted by considering all environmental and ecological aspects. International cooperation will be necessary to support this kind of agricultural development in Korea, China, and Japan. In our future work, we will attempt to develop assessment strategies for environmental factors using ecological statistics based on biodiversity data and non-point monitoring data.

Agro-climatological backgrounds for impact assessment in East Asia and speculation about future resources

Yousay Hayashi

National Institute for Agro-Environmental Sciences, 3-1-3 Kannondai,
Tsukuba, Ibaraki 305-8604, Japan

Abstract

Over the past 40 years, the population of the world has doubled, but the quantity of water consumed by people has increased at twice the rate of population growth. Today, humans use about half of the world's usable fresh water, and water resources that are essential for the production of food have been affected. Because the current world population of ca. 6.3 billion is anticipated to increase by 50 % by 2050, and economic development is expected to accelerate, the consumption of water will increase drastically in East Asia, and especially in China, South Korea, and Japan. This paper outlines agro-climatological characteristics peculiar to the water regime in Monsoon Asia, and discusses the importance of phenomena that help in understanding the problems of the agricultural environments in China, South Korea, and Japan.

Keywords: agro-climatology, East Asia, agricultural resources, global warming, impact assessment

1. Introduction

When we consider environmental change, human population increases are one of the most serious problems. The world had an estimated population of 5 to 15 million long before the Christian era, and mankind lived mainly as hunter-gatherers. However, populations increased significantly after the development of agrarian civilizations, reached about 500 million by 1650 (prior to the Industrial Revolution), and exceeded 1 billion by around 1800. Thereafter, the world population increased to 2.5 billion by 1950, and reached 6.1 billion by 2000. This means that the population has increased by as much as 140% during the past 50 years.

Figure 1 (a) shows the distribution of the world's population in recent years, and indicates that East Asia represents one of the most densely populated areas in the world. The intense temperature rise that has been observed over the past 100 years in the same region (Fig. 1b)

(a) Population desity

(b) Annual temperature trends, 1901 to 2000

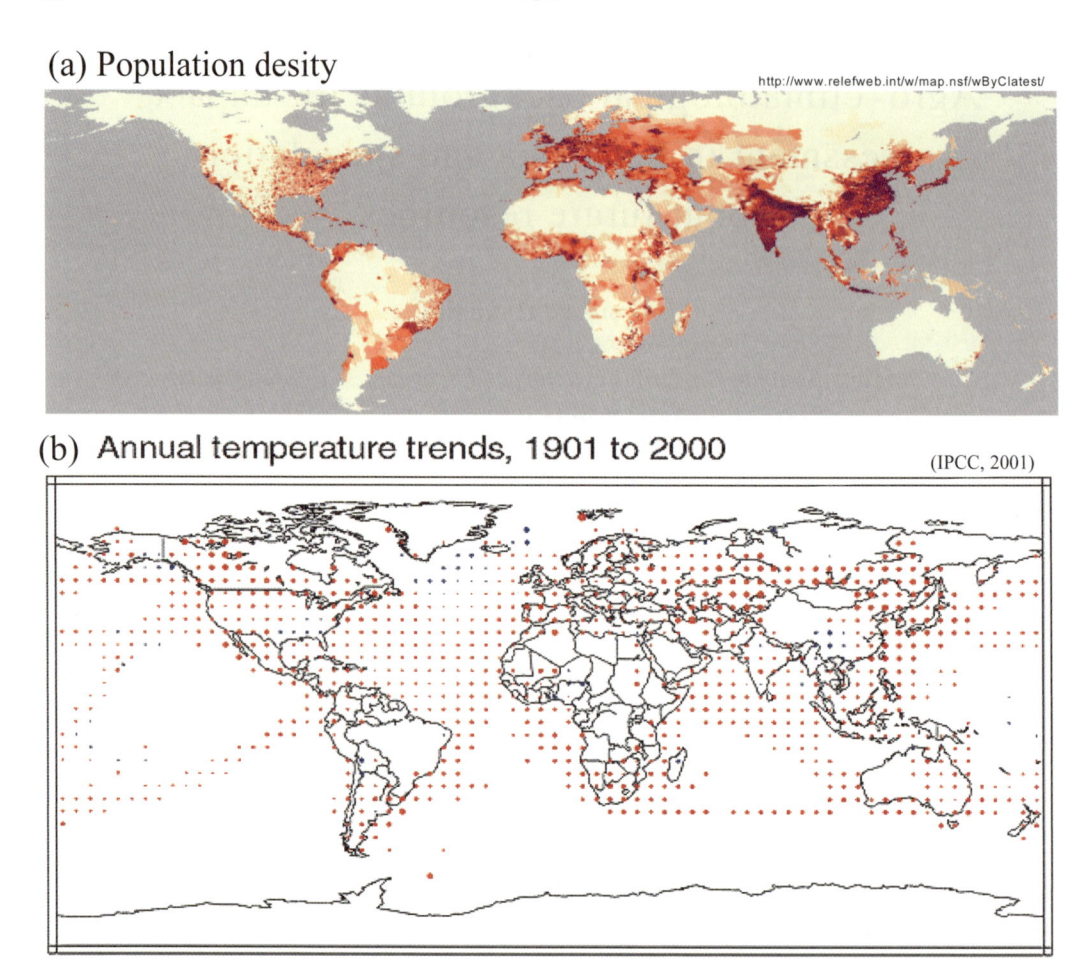

(IPCC, 2001)

Fig. 1 World distribution of population density(a) and annual temperature trends

bears eloquent testimony to the present regional distribution of population density. IPCC (1996, 2001) has suggested that the spatial distribution of surface temperatures shown in Fig. 1 (b) are anomalous. The red and blue circles indicate areas with positive trends (increasing temperatures) and negative trends, respectively. The size of each circle reflects the magnitude of the trend. Most warming is observed over mid- and high latitudes in Asia, including China, South Korea, and Japan.

The world's average air temperature and precipitation over land areas have increased by 0.6 ℃ and decreased by 0.7 %, respectively, over the past 100 years. Comparing these changes among various parts of the globe has shown that northern Asia, including China, South Korea, and Japan, experienced the greatest increases (1 ℃) in average temperature and the greatest variability (12.1%) in precipitation. If this trend continues, agriculture in and around East Asia may be first to suffer the negative effects of long-range trends. Given this background of climate change, the growing population in the region, and the decrease in cultivation of farmland due to changes in land use, the importance of predicting the state of future agriculture in East Asia is becoming critical.

2. Agro-environmental factors in the three countries

China, South Korea, and Japan all share the problems faced by agricultural environments throughout East Asia. The basic elements essential for comprehending the conditions in these countries, such as population, land area, quantity of chemical fertilizers, and precipitation, are shown in Table 1 together with Asian and world averages.

The ratio of crop land area to total national area is 13.3% in China, 17.4% in South Korea, and 12.1% in Japan; all three values are greater than the world average. All three countries use large quantities of chemical fertilizers per unit of crop land area. In particular, South Korea has a tendency to use large amounts of fertilizers, up to 500 kg/ha. Though China has the lowest mean amount of rainfall (660 mm per year; Table 1), its total rainfall is about ten times that of Japan because of its larger land area. China and Japan have the same amount of rainfall per person (about 5000 m^3 per year), but South Korea has only about half this per-capita rainfall. In East Asia, the quantity of chemical fertilizers used over crop land (Table 1) suggest that manure is used heavily in order to increase crop yield. Because precipitation levels are high in East Asia, agriculture is highly productive. However, this heavy use of fertilizer also increases the risk that chemicals from the fertilizer flowing out of cultivated lands will have a greater impact on the environment than in other regions.

Table 1 Backgrounds of Agro-Environmental Conditions in East Asia

Element	Unit	China	South Korea	Japan	Asia	World
Population (1999)	1000 person	1,266,838	46,858	126,505	3,634,000	5,978,000
Total Land Area (1)	1000ha	932,641	9,873	37,652	3,085,414	13,048,407
Crop Land Area (2)	1000ha	124,145	1,722	4,569	498,519	1,380,848
(2) / (1)	%	13.3	17.4	12.1	16.2	10.6
Chemical Fertilizer Nitrogen	1000t	22,446	451	476	47,336	82,421
(FAOSTAT) Phosphate	1000t	9,185	188	476	17,954	32,912
Potassium	1000t	3,447	235	382	7,631	22,022
Total	1000t	35,078	874	1,334	72,921	137,355
Amount of Chemical Fertilizer per capita	g/person	27.7	18.7	10.5	20.1	23.0
Chemical Fertilizer Use over Crop Land	t/ha	0.283	0.508	0.292	0.146	–
Average Precipitation over Land Area	mm/year	660	1,143	1,714	–	–
Total rain fall over Land Area	billion m^3/year	6,334	113	648	–	132,000
Amount of Precipitation per capita	m^3/person/year	5,000	2,412	5,122	–	22,081

Data were partly quoted from the bilateral research between Japan and Korea titled "Climate variability and its impacts on paddy rice production in Japan and South Korea"

The total annual precipitation for the entire Earth amounts to $5.05 \times 10^{14}\,m^3$ (equivalent to an average precipitation of about 970 mm/year). As a matter of course, the same amount ($5.05 \times 10^{14}\,m^3$) enters the atmosphere each year via evaporation from water surfaces and via plant transpiration. Because the amount of water in the air at any one time is $0.16 \times 10^{14}\,m^3$, water in the atmosphere is replaced about every 11.2 days through rainfall and evaporation. If the water present in the atmosphere would be lost within 12 days without this resupply of water, this suggests that water circulates *rapidly* between the Earth's surface and the atmosphere. Because of its higher-than-average rainfall, East Asia seems to be located in an area that has particularly rapid circulation of water.

3. Balance between global demand for and supply of cereal crops

The changes in the harvested area of cereals, production of cereals, yield of cereals per unit area, population, yield of cereals per person, and harvested area of cereals per person during the past 40 years are shown in Table 2. World cereal production, which directly affects the overall availability of food, increased by around 225% from an average of 920×10^6 t (1961 to 1963) to 2070×10^6 t (1999 to 2001). Because the world population only doubled during the same period, the cereal yield per person increased by 14%, from 297 to 339 kg/person. The improvement in the overall global food situation has resulted from the production of cereals continuing to increase faster than the population.

However, because the population has increased continuously without a parallel increase in the area of farmland around the world, the mean harvested area per person has consistently decreased. The changes in global production of cereals and in harvested area per person (Table 2) indicate that the harvested area of cereals has decreased despite the increased production of cereals. The harvested area per person has declined by nearly 50%, from 20.8 a (1961 to 1963) to 11.1 a (1999 to 2001).

In general, the continuing increase in the yield of cereals exceeded the rate of population increase, and until 1980, a sufficient supply of cereals could be provided without increasing the area harvested. Since the 1990s, however, the rate of yield increase has fallen below the rate of population increase; as a result, the overall global demand for cereals may soon exceed the supply. As shown in Table 3, the deceleration of the rate of increase in the yield of cereals is obvious.

Table 2 Changes in yield of cereals, harvested area per person and the like in recent years

	1961–1963	1999–2001
Harvested area of cereals (hundred million ha)	6.5	6.7
Production of cereals (hundred million tons)	9.2	20.7
Yield of cereals per unit area (tons/ha)	1.4	31.
Population (hundred million persons)	31	61
Yield of cereals per person (kg/person)	297	339
Harvested area of cereals per person (are/person)	20.8	11.1

(FAOSTAT)

Table 3 Changes in yield of cereals per unit area and in the rate of its increase

	1961–1963	1971–1973	1981–1983	1991–1993	1999–2001
Yield per unit area (t/ha)	1.41	1.9	2.31	2.74	3.09
Rate of increase (%)		3.0	2.0	1.7	1.5

(FAOSTAT)

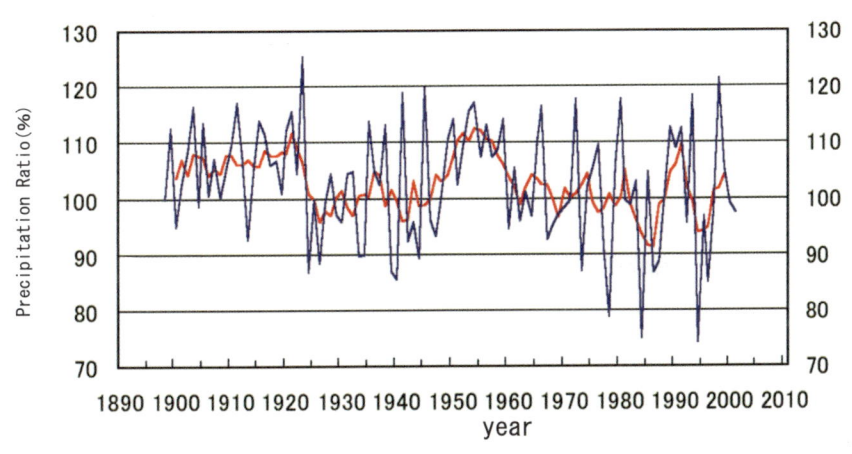

Fig. 2 Yearly change of amount ratio in precipitation of Japan(JMA,2001)

4. Fluctuation of precipitation in East Asia

Although the demand for water resources is increasing, the amount of precipitation (the main source of this water) has fluctuated significantly in recent years (Fig. 2). Most significantly, the observed variation has increased greatly, with at least three periods of dramatically lower precipitation. This phenomenon relates closely to the activity of the Asian Monsoon, which is the primary mechanism for transporting water vapor over China, South Korea, and Japan. In consequence, crop production in East Asia varies widely and may decrease significantly from its potential level if this trend continues in the future.

China uses an index called "damaged area" to indicate the degree of the damage caused by natural disasters such as droughts or flooding. A "damaged area" is defined as an agricultural area that sustains a reduction in yield of at least 30%. The total damaged area caused by droughts and floods shows an increasing trend over the past 30 years, accompanied by greater yearly fluctuations. The annual change in the total damaged area indicated a prominent peak in 1994, when food crop production decreased by 2% to 3% compared with the previous year (Hayashi, 2000).

There were clear characteristic regional distributions of damage by flood and by drought in 1994. Figures 3 and 4 show the proportion of each Chinese province's total planted area

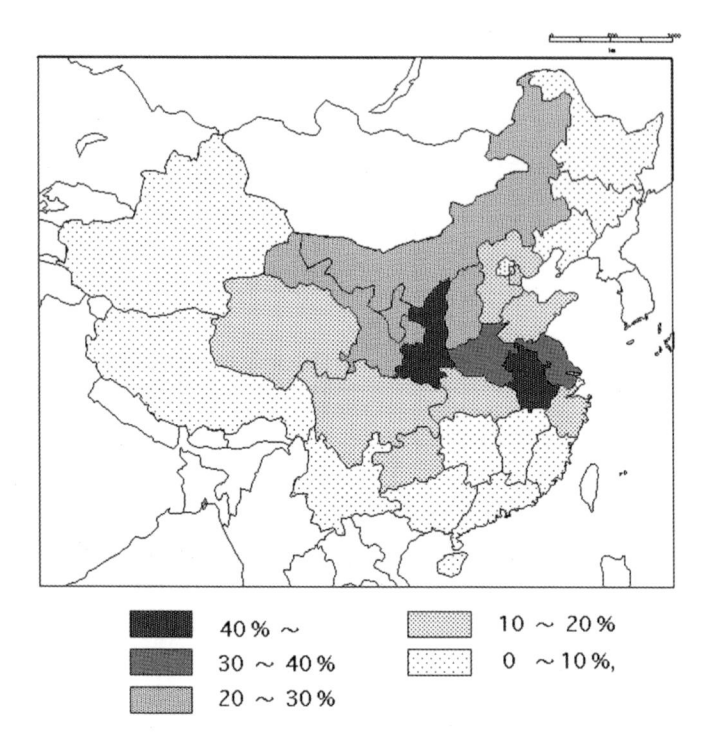

Fig. 3 Flooding Area over China in 1994

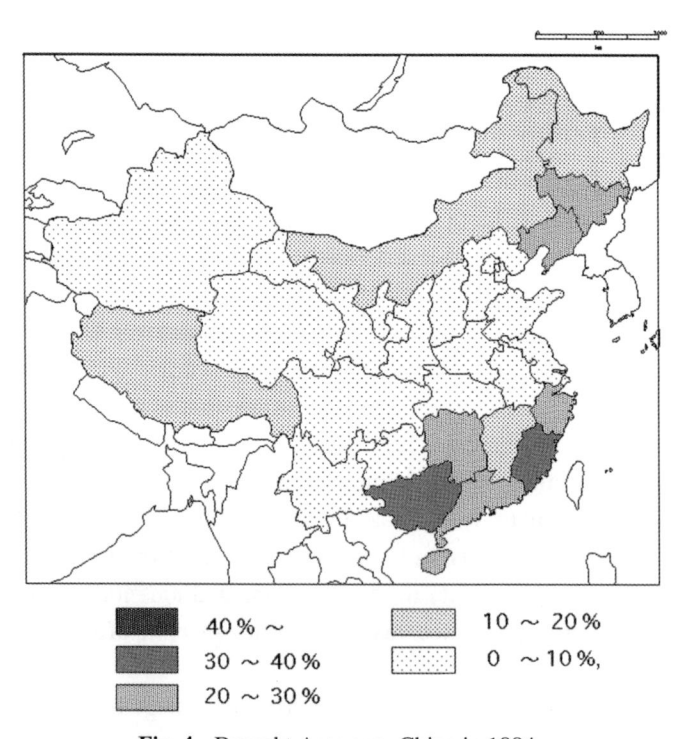

Fig. 4 Drought Area over China in 1994

damaged by flooding and drought, respectively. Provinces such as Jiangsu, Anhui, Henan, and Shanxi, north of the Yangtze River, suffered the most from flooding. Conversely, provinces such as Zhejiang, Fujian, Guangdong, Hunan, and Jiangxi, situated south of the Yangtze River, suffered most from drought. The concentration of these damages in specific areas is considered to be closely related to the temporal and geographical scales of these climatic disturbances.

5. Traditional pessimism and optimism towards "the food problem"

5.1. Pessimism

Changes in the global environment have highlighted two conflicting opinions related to the problem of future food supplies (Hayashi, 2001). The pessimistic attitude has its origin in Malthusianism. Thomas Malthus studied both population growth and the growth in food production in 18th century Europe; he published *An essay on the principle of population* in 1798. In this study, Malthus drew a famous conclusion: that population growth, if unchecked, increases geometrically, whereas the food supply for humans increases only arithmetically, leading inevitably to starvation and population collapses as mankind exceeds its food supply. This conclusion was drawn from observations that harvests tend to increase linearly, that population tends to increase exponentially, and that the world's area of arable land is fixed. Malthus's pessimistic prediction did not come true, because the area of cultivated land increased dramatically when the New World (North and South America) became available to European farmers. However, current predictions are that the world's population will reach 10 billion in the near future, and there is little room for further agricultural development; consequently, Malthus's prediction looks increasingly realistic.

More recent pessimism has been described in works such as *The Limits to Growth* by Meadows. The authors pointed out the possibility of mankind confronting serious restrictions on growth dur to the energy crisis that was then affecting the world. However, the scenario predicted by *The Limits to Growth* was that the world would experience a desperate shortage of farmland long before the present, and is thus not considered to be realistic. In particular, the area of farmland required to feed one person in *The Limits to Growth* (0.4 ha) proved to be an overestimation, at least on the global scale. However, it is true that a farmland area of 0.9 ha is required to feed a single person with North American dietary habits.

Lester Brown is a famous pessimist who formerly led the Worldwatch Institute and who founded the Earth Policy Institute (a nongovernmental organization) in 2003. Brown recently published *Plan B* (Brown, 2003). Brown believes that the present system, with a decreasing area of cultivated land balanced by increasing yields per unit area, will collapse in the near future. His opinion is unique in that he stresses the insufficient availability of agricultural water as the factor that most limits the expansion of agricultural land. Whether stable harvests can be secured in the future can only be discussed in the context of the deteriorating natural environment, as represented by abnormal climatic conditions, climatic changes, depletion of underground water, and salinization of agricultural land.

5.2. Optimism

In sharp contrast to the pessimistic opinions presented above, the International Food Policy Research Institute (IFPRI) presents a more optimistic opinion. IFPRI was established in 1975 to help solve the food problems of developing countries, and has its headquarters in Washington D.C. The forecasting methods used by the World Bank and the Food and Agriculture Organization (FAO) of the United Nations are essentially the same as IFPRI's.

The basic concept behind the simulation models used in these studies is based on econometrics. In this approach, prices are determined by balances between supply and demand, which, in turn, determine the future production of all products. The estimates produced by the IFPRI model suggest that the prices of grains and meats will be about 20% of its current level in 2020, and that the incremental demands in developing countries can thus be easily met by exports from developed countries. IFPRI is also optimistic about such issues as the extent of the degradation of soils, the rate of global environmental deterioration, and the rate of disappearance of farmland.

5.3 Pessimism or optimism?

In general, pessimistic opinions are voiced by researchers with scientific and technological backgrounds, whereas optimistic opinions are voiced by economists. When Meadows et al. (1972) published *The Limits to Growth*, the combined effects of the poor crop yields in the United States and the high prices of energy inputs suggested a dim future for agricultural production. Against such a background, food price indices remained high. In reality, the supply and demand situation for food has never become as tight as the pessimists predicted and food price indices have consistently dropped since the 1980s. This situation suggests to the economists that food production will be increased on a global scale if food prices rise, and that production can easily catch up with demand. These economists believe that declining prices for various measures of production and transportation of goods (including food), combined with the unification of the world's economies that has been brought about by termination of the Cold War, are making the world more economically efficient. Their conclusion is that food production will respond efficiently to changes in demand and that price inflation is disappearing as a significant force. However, even if the econometric models are correct, economists have been unable to dismiss the valid concerns expressed by pessimistic researchers who express increasing concern over climate change and other potential environmental problems.

Conclusions

The combined population of China, South Korea, and Japan amounts to 25% of the world's population, with China accounting for 86% of this total. As populations increase and land use changes, characterized by a decrease in the area under cultivation, the proportion of the cultivated area that will suffer from climatic damage is increasing. In the past few decades, East Asia, in particular, has suffered from the greatest drought and flood damage in recent history. These phenomena demonstrate the vulnerability of agriculture to global environmental changes. In impact-assessment studies, it is becoming increasingly important to focus on the variations in yearly climatic conditions from normal conditions and the unique characteristics of specific regions, in addition to such long-range trends in environmental conditions as global warming.

It is generally said that about 1.4×10^{15} m^3 of water exists on the Earth. About 97.5% of this water is sea water, and the remaining 2.5% is fresh water. Most of the fresh water exists as ice in the northern and southern polar regions, and liquid fresh water, which exists as river or lake water and groundwater, accounts for only about 0.8% of the total water on Earth. In addition, river water and lake water, which are relatively easy to access, account for only 0.01% of this total. Agricultural activities are fed by this unexpectedly tiny amount of water. Given the problems that face this precious resource, the pessimists may have some justification in believing that Earth's capacity to sustain exploitation of its water resources may have reached its limit.

References

Brown, L.R. (2003) Plan B. (ed. by K. Hojo) Worldwatch-japan : 348 pp

FAOSTAT (2004) Agricultural database, http://faostat.fao.org/faostat/collections/subset = agriculture

Hayashi, Y. (2000) Effect of global environmental change on agriculture in East Asia-Background of agricultural vulnerability. Text for Development of National Inventories and Strategies against Climate Change, TBC JR, 00-125, Japan International Cooperation Agency: 16 pp

Hayashi, Y. (2001) Vulnerability assessment and adaptation measures (Agriculture). Text for Development of National Inventories and Strategies against Climate Change, TBC JR, 01-127, Japan International Cooperation Agency: 13pp

IPCC. (1996) Climate Change 1995-Impacts, Adaptations and Mitigation of Climate Change: Scientific-Technical Analyses. Cambridge University Press: 878 pp

IPCC (2001) Summary of Policymakers-A report of Working Group I of the Intergovernmental Panel on Climate Change. Intergovernmental Panel on Climate Change, http://www.grida.no/climate/ipcc_tar/

Japan Meteorological Agency (ed.) (1995) Global Warming Monitoring Report 1994. Japan Meteorological Agency: 138 pp

Japan Meteorological Agency (ed.). (2000) Report of Climate Change Monitoring. Japan Meteorological Agency: 55 pp

Medouws, D.H., Medouws, D.L., Randers, J. and Behrens III, W.W. (1972) The limits to growth. (trans. by S. Okita) Daiamond-Sha: 203 pp

Multhus, T.R. (1973) An essay on the principle of population. (trans. by Y. Nagai) Chuokoron-shinsha Inc.: 242 pp

Options for mitigating CH_4 emissions from rice fields in China

Zucong Cai and Xu Hua

*Institute of Soil Science, Chinese Academy of Sciences,
Nanjing 210008, China*

Abstract

Rice fields are an important source of atmospheric CH_4, a greenhouse gas whose levels have increased continuously since the Industrial Revolution. Options for mitigating CH_4 emissions from rice fields have been assessed intensively, even though their necessity is debated. In this paper, we summarize field measurements and greenhouse experiments, and discuss the options for mitigating CH_4 emissions from rice fields in China. There is a high potential for reducing CH_4 emissions from permanently flooded rice fields, which are mainly distributed in southern and southwestern China, during the fallow season. Draining floodwater from these fields during the fallow season stops CH_4 emissions both during that season and during the following growing season. However, this simple solution is not applicable for most permanently flooded fields. An alternative involves ridge-cultivation of the permanently flooded fields, which could reduce annual CH_4 emissions by 26.6% during the growing season and by 40.0% during the fallow season. For temporarily flooded rice fields, an exponential relationship between soil moisture during the fallow season and CH_4 emissions during the growing season suggested that keeping soils as dry as possible during the fallow season would also reduce CH_4 emissions during the growing season. Incorporation of organic manure and crop residues into soils during the drained season instead of immediately before rice transplanting could significantly decrease the stimulation of CH_4 emissions.

Keywords : rice, methane emission, permanently flooded rice field, soil moisture, fallow season

1. Introduction

Since rice fields are considered to be an important source of atmospheric CH_4, options for mitigating CH_4 emissions from rice fields have been intensively assessed (Inubushi et al., 1997; Yagi et al., 1997; Wassmann et al., 2000; Aulakh et al., 2002). Several options have been demonstrated to be effective for mitigating CH_4 emissions from these fields, for instance, intermittent irrigation during the growing season (Wassmann et al., 1993; Yagi et al., 1997), application of compost or biogas residues instead of fresh organic matter such as straw and green manure (Shin et al., 1996; Wassmann et al., 2000; Zheng et al., 2000), application of organic manure during the fallow season to allow aerobic decomposition of the manure before it is submerged (Shin et al., 1996; Yagi et al., 1997), application of sulfur-containing nitrogen fertilizers instead of urea and ammonium bicarbonate (Inubushi et al., 1997), and selection of appropriate rice cultivars (Aulakh et al., 2002). In summarizing field measurements conducted in China, India, Indonesia, Thailand, and the Philippines, Wassmann et al. (2000) found that in comparison with prevailing practices, optimizing irrigation patterns by providing additional drainage periods or an early mid-season drainage reduced CH_4 emissions by 20% to 93%. In comparison with baseline practices using high quantities of organic soil amendments, the use of compost, biogas residues, and direct wet seeding reduced CH_4 emissions by 37% to 42%, 84% to 90%, and 78% to 84%, respectively. In comparison with baseline practices using prilled urea as the sole N source, the use of ammonium sulfate reduced CH_4 emissions by 10% to 67%. In all rice ecosystems, CH_4 emissions could be reduced by fallow incorporation (11%) and mulching (11%) of rice straw as well as by the addition of phosphogypsum (9% to 73%).

However, if the mitigation of CH_4 emissions from rice fields is necessary—something that is still being debated—the formidable obstacles to adoption of these mitigation options should be taken into account (Yagi et al., 1997). For example, intermittent irrigation would increase the costs for multiple irrigation, would require that sufficient water sources be available, and might stimulate N_2O emissions (Cai et al., 1997, 1999). Furthermore, the strategies chosen should be compatible with local rice cultivation practices, soil types, and so on. In the present paper, we discuss the highest-potential rice fields and times at which interventions could reduce CH_4 emissions in China, based mainly on our research findings.

2. High-potential rice fields for mitigating CH_4 emissions in China

Methane emissions from rice fields in China have been a serious concern because the area of rice cultivation and the amount of rice production accounted for up to 22% and 36%, respectively, of the world totals. Field measurements conducted in the late 1980s and early 1990s showed that CH_4 fluxes from Chinese rice fields were much higher than those measured elsewhere in the world. Based on very limited field measurements of CH_4 emissions, early estimates of annual CH_4 emissions from Chinese rice fields were up to 30 Tg CH_4/yr (Khalil et al., 1991) or even higher (Kern et al., 1997). As a result of these early reports and the importance of ongoing increases in atmospheric CH_4 concentrations in the greenhouse effect, the CH_4 emissions from rice fields in China attracted the attention of scientists around the world. Great efforts have been made to estimate CH_4 emissions from these rice fields (e.g., Kern et al., 1997; Huang et al., 1998; Sass et al., 1999; Matthews et al., 2000; Verburg et al., 2001; Li et al., 2002). With the accumulation of field measurements, we now know that there are large spatial and annual variations in CH_4 emissions from these fields. The reported CH_4 emissions during the growing season ranged from 0.3 to 205 g CH_4/m^2 (Cai et al., 2000), and these emissions fell within the range of values measured elsewhere in

Table 1 CH$_4$ emissions from rice fields during rice growing period affected by water status in the preceding non-rice growing season (Kang et al., 2002)

Site	Latitude/Longtitude	Seasonal CH$_4$ emission (g CH$_4$ m^{-2})		
		Drained (A)	Flooded (B)	A/B
Guangzhou	23° 15′ N/113° 6′ E	11.8 ± 11.7	76.0 ± 35.4	0.21
Yingtan	28° 12′ N/117° 6′ E	90.4 ± 39.4	158.5 ± 65.8	0.57
Changsha	28° 9′ N/113° 6′ E	55.4 ± 32.0	103.5 ± 34.6	0.53
Chongqing	29° 48′ N/106° 18′ E	47.8 ± 24.3	56.1 ± 22.3	0.85

the world; however, a few were unusually large.

Rice-based ecosystems in China have two main seasons: the summer rice-growing season and the winter fallow season. The former starts from late March to mid-June and ends from late August to November, and varies both from southern to northern China and with the number of rice crops (double and single). The later date is between the two seasons of rice cropping. The water regimes of Chinese rice fields differ substantially during the fallow season, and can be divided into two types: flooding and draining. Rice fields that are drained during the fallow season are either planted with upland winter crops or are left fallow under drained conditions. Upland winter crops planted in rice-based ecosystems are commonly winter wheat, oilseed rape, and various forms of green manure. The fields that are flooded during the fallow season are usually left permanently flooded and are not planted during that season.

Summarization of our field measurements showed that compared with winter-drained rice fields at the same location, winter-flooded rice fields emitted considerably more CH$_4$ during the growing season (Table 1). Our early measurements were mainly conducted in winter-flooded rice fields, and the results represented CH$_4$ emissions only from this kind of field. Recent estimates of CH$_4$ emissions from rice fields have been much smaller than the early estimates, ranging from 3.73 to 15 Tg CH$_4$ (Kern et al., 1997; Huang et al., 1998; Sass et al., 1999; Matthews et al., 2000; Verburg et al., 2001; Li et al., 2002).

Flooding is a prerequisite for CH$_4$ production. Flooding during the fallow season maintains anaerobic conditions, which extends CH$_4$ emissions between growing seasons. Measurements carried out in a lysimeter experiment in Japan showed that CH$_4$ emissions during the fallow season were equivalent to between 14% and 18% of emissions during the previous growing season if the paddies were continuously flooded all year round (Yagi et al., 1998). However, the magnitudes of CH$_4$ emission measured in a permanently flooded rice field in China during the fallow season were much higher than in the lysimeter experiment and were even larger than CH$_4$ emissions during the growing season from most rice fields, which were well-drained during the preceding fallow season (Cai et al., 2003). Our 6-year field measurements carried out in Chongqing, China, showed that although the mean CH$_4$ flux throughout the fallow season was much lower than that during the growing season, the total CH$_4$ emission during the fallow season (53.5 g CH$_4$ m^{-2}) was very close to that (67.2 g CH$_4$ m^{-2}) during the growing season because the length of the fallow season was more than twice the length of the growing season (Cai et al., 2003).

Though there are no reliable statistics on the area of permanently flooded rice fields, which are mainly distributed in southern and southwestern China, statistics from the Second National Soil Survey showed that the area of Gleyic Paddy Soils, which were believed to represent permanently flooded rice fields, was 2.52 million ha (Office for the National Soil Survey, 1997). However, non-Gleyic Paddy Soils may also be flooded during the fallow season. Hence, the total area of winter-flooded rice fields would be greater than the area of Gleyic Paddy Soils. According to a report on paddy soils in China, there were 2.7 to 4.0 million ha of permanently flooded rice fields (Lee, 1992). Therefore, it is clear that winter-flooded (or permanently flooded) rice fields represent a substantial opportunity for mitigating CH_4 emissions.

For this reason, we have made the task of mitigating CH_4 emissions from winter-flooded (or permanently flooded) rice fields a high priority because of their large area and their substantial CH_4 emissions compared with fields that are drained during the fallow season.

The key factor driving high CH_4 emissions from permanently flooded rice fields is flooding during the fallow season. A simple way to reduce CH_4 emissions from these fields would thus be to drain floodwater completely and generate aerobic conditions during the fallow season. Our results showed that compared with the 6-year average annual CH_4 emission from a permanently flooded rice field, draining floodwater and planting a winter crop during the fallow season could reduce average annual CH_4 emission by 67.8%; stopping CH_4 emissions during the fallow season by generating aerobic conditions through drainage of floodwater accounted for 44.2% of this total (Cai et al., 2003).

Unfortunately, draining floodwaters is not applicable for most permanently flooded rice fields. Several factors lead farmers to maintain floodwater in their fields all year round: (1) Drainage conditions are too poor to completely drain floodwater from the soil because of the depressed topography or a high water table. (2) The irrigation system is poorly developed and depends to a greater or lesser extent on rainfall. If the fields were drained in winter, and spring precipitation were not sufficient, the fields could not be flooded in time to transplant rice for the new growing season. (3) Farmers retain the floodwater layer after rice harvesting without considering why they do so. Only in this last case would completely draining the floodwater layer to generate aerobic conditions during the fallow season be easily applicable. In the first two cases, completely draining the floodwater layer during the fallow season would be applicable only where well-developed drainage and irrigation systems exist and water supply is good. Nevertheless, instead of completely draining floodwater during the fallow season, a relatively new approach called ridge-cultivation would still permit mitigation of CH_4 emissions from permanently flooded fields.

Ridge-cultivation of these fields was developed about two decades ago (Xie, 1988). In ridge-cultivation, 30-cm-wide fixed ridges are constructed (Fig. 1). Rice plants grow on both sides of the ridge. Ditches between ridges are filled with water to a level up to 3 cm below the top of the ridges during the growing season. After harvesting, the field is left fallow and the water level in the ditches is allowed to drop to between 5 and 10 cm below the ridge

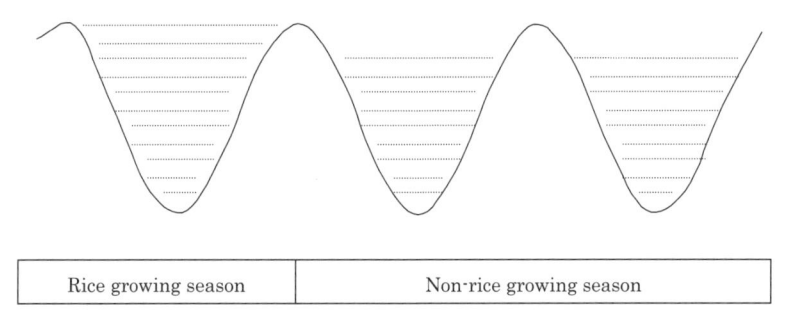

Fig. 1 Scheme of ridge-cultivation for permanently flooded rice field. Ditches between two ridges are filled with water to the level of 0-3 cm below the top of the ridge in RGS and to 5-10 cm in the non-RGS

tops. No tillage is practiced in ridge-cultivation, but mud accumulated in the ditches during the previous year is used to cover weeds that have become established during the fallow season and stubble remaining from the previous rice plants. Through this practice, the ridges can be maintained for a long time. Our 6-year measurements in Chongqing, China, showed that compared with traditional cultivation of permanently flooded rice fields, CH_4 emission from fields with ridge-cultivation decreased, on average, from 67.2 to 49.3 g CH_4/m^2 during the growing season (a 26.6% reduction) and from 53.5 to 32.0 g CH_4/m^2 (a 40.0% reduction) during the fallow season (Cai et al., 2003).

3. Optimal time for mitigating CH_4 emissions from rice fields in China

Previous assessments of the alternatives for mitigating CH_4 emissions from rice fields focused mainly on the growing season; these alternatives included intermittent irrigation, the use of special rice cultivars, and fertilization. Intermittent irrigation is an effective option. Summarizing the available data on CH_4 emissions from Chinese rice fields (Table 2), we found that on average, intermittent irrigation did not effectively mitigate CH_4 emissions early in the growing season, but was effective during the middle and late growing season. The mechanisms responsible for this difference are currently unknown. In fact, intermittent irrigation has already been commonly adopted in China because the practice effectively controls the number of tillers and prevents the soil from becoming excessively anaerobic.

Table 2 Comparisons of CH_4 emissions between intermittent flooding (IF) and continuous flooding (CF) during rice growing period averaged over available data in China

Rice season	Mean CH_4 flux (mg CH_4 m^{-2}h^{-1})		
	IF	CF	IF/CF, %
Early rice	10.9	10.1	108
Middle rice	5.3	12.8	41
Late rice	12.4	19.2	65

Therefore, the potential for mitigating CH_4 emissions by further extending the area of rice fields cultivated in this manner is limited. However, there is still a high potential for mitigating CH_4 emissions by managing soil water status during the fallow season.

It is well known that CH_4 is the product of reduction at the end of the oxidation − reduction series in soils. If a rice field is flooded year-round, the soil retains an anaerobic status and its oxidation − reduction potential (soil Eh) is usually sufficiently low for CH_4 production. Temperature and the presence of suitable organic substrates then become the main factors that control the rate of CH_4 production. For a field that is drained during the fallow season, soil Eh decreases gradually after flooding to permit the transplanting of rice seedlings, and there is a certain delay before the Eh value becomes sufficiently low to permit CH_4 production (see Fig. 2). In response to this pattern of changes in soil Eh, CH_4 emission after rice transplanting is similarly delayed compared to emissions from rice fields flooded during the preceding fallow season (Fig. 3). In such cases, soil Eh is the dominant factor controlling CH_4 emission (Xu et al., 2002). Continuous flooding during the fallow season also preserves favorable conditions for the survival of methanogenic bacteria. If a rice field is

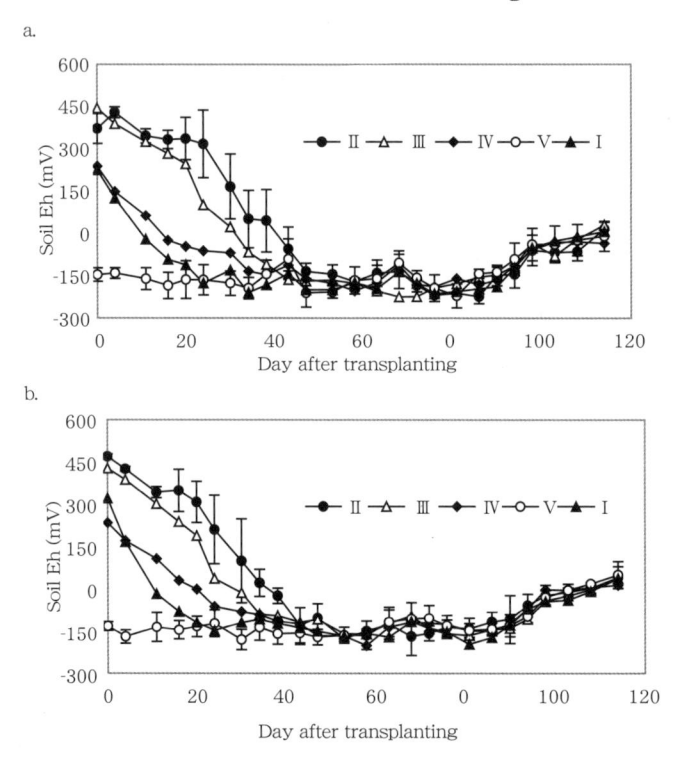

Fig. 2 Time courses of soil Eh change after rice transplanting affected by soil moisture in preceding non-rice growing season. a, paddy soil collected from Wuxi, Jiangsu province; b, paddy soil collected from Yingtan, Jaingxi province. I, II, III, IV and V represent soil moisture of air-dry, 25-35%, 50-60%, 75-85%, and 107% (flooded) of field water capacity in the preceding non-rice growing season, respectively (cited from Xu et al., 2002).

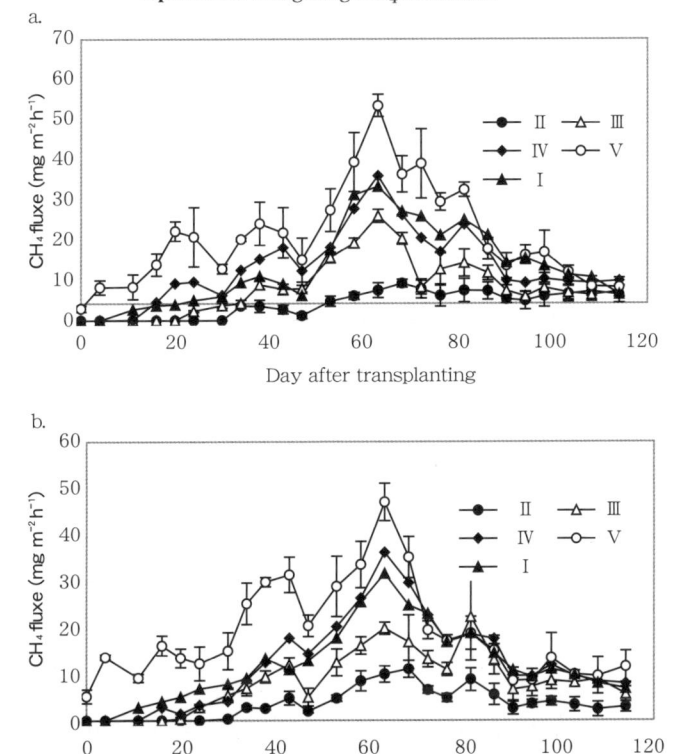

Fig. 3 Seasonal variation of CH_4 fluxes during the rice growing period affected by soil moisture in the preceding non-rice growing season. a, paddy soil collected from Wuxi, Jiangsu province; b, paddy soil collected from Yingtan, Jaingxi province. I, II, III, IV and V represent soil moisture of air-dry, 25-35%, 50-60%, 75-85%, and 107% (flooded) of field water capacity in the preceding non-rice growing season, respectively (cited from Xu et al., 2002).

drained during the fallow season, oxygen diffuses into the soil and decreases populations of methanogenic bacteria or depresses their activity (Ueki et al., 1997). The depressive effect of drainage on methanogenic bacteria during the fallow season would also reduce CH_4 emissions during the following growing season, since populations of these bacteria take some time to reach their formerly high levels.

Continuous flooding during the fallow season is an extreme water regime. However, the soil moisture in rice fields during the fallow season varies with precipitation, topography, and type of management. In China, the soils of rice fields generally become wetter moving from the north to the south. Our results have demonstrated that soil moisture during the fallow season plays an important role in controlling CH_4 emissions during the following growing season, suggesting that CH_4 emissions during the fallow season are likely to be greater in the south than in the north. A pot experiment showed that with increasing soil moisture during the fallow season, the mean CH_4 flux during the following growing season also increased, except for pots allowed to air-dry during the fallow season (Xu et al., 2002). Spatial variations in CH_4 emissions from temporarily flooded rice fields in China are predominantly

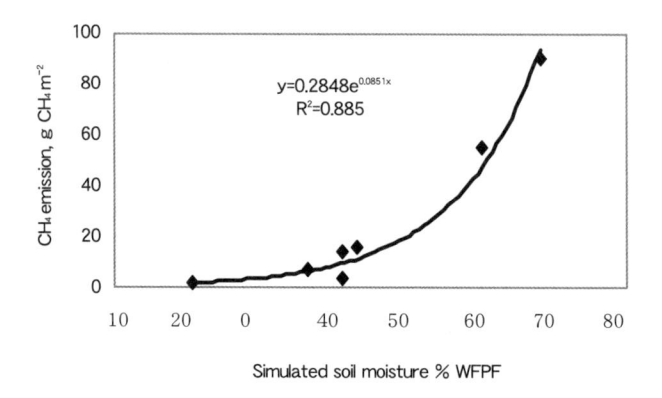

Fig. 4 Relationship between soil moisture in the non-RGS and CH_4 emission during RGS (Kang et al., 2002)

controlled by the amount of precipitation during the fallow season (here, defined as extending from Nov. 1 to March 31). Results from a model that simulates soil moisture changes during the fallow season were significantly related to annual CH_4 emissions from temporarily flooded rice fields (Fig. 4). Therefore, keeping soil as dry as possible during the fallow season and preventing waterlogging through good management would reduce CH_4 emissions during the following growing season. It has also been demonstrated that CH_4 emissions during the growing season could be reduced by using underground drainage to reduce soil moisture during the fallow season (H. Tsuruta, personal communication). The exponential relationship in Fig. 4 suggests that reducing soil moisture would mitigate CH_4 emissions more effectively in wet soil than in dry soil.

Adjusting crop rotations to allow the development of a soil's aerobic status could also effectively reduce CH_4 emissions during the growing season. The traditional crop rotation in Guangzhou is two rice crops followed by planting of a winter upland crop. As a result of economic development in the region, the rice crop has been partially replaced by vegetables. During the early rice-growing season in 1995, we measured CH_4 emissions in three types of plot: one in which double rice crops were followed by an upland crop (traditional crop rotation), one in which early-season rice was followed by two upland crops, and one in which only upland crops were grown. The water regimes, fertilization rate and type, and other agricultural practices were the same in all three plots during the early rice-growing period. In the baseline (traditional) crop rotation, CH_4 emissions decreased by 73% compared with the plot with two consecutive upland crops and by 92% compared with the plot that contained only upland crops (Cai et al., 2000).

4. Time of application of crop residues and organic manure

The application of organic manures or crop straw is encouraged because it maintains soil organic matter content and soil fertility. However, the practice stimulates CH_4 emission from the rice field if organic manure or crop straw is applied immediately before rice transplanting.

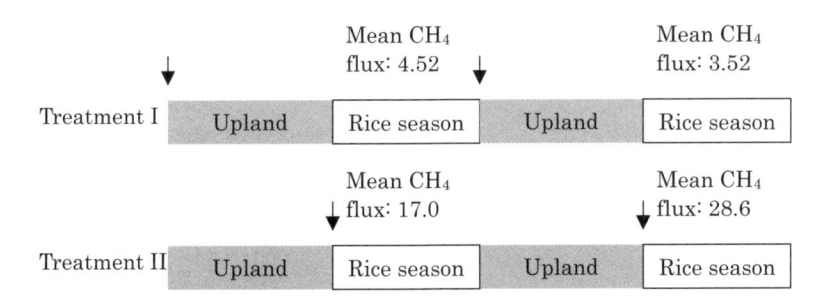

```
                        Mean CH₄                  Mean CH₄
                        flux: 4.52 ↓              flux: 3.52 ↓
              ↓
Treatment I   | Upland | Rice season | Upland | Rice season |

                        Mean CH₄                  Mean CH₄
                      ↓ flux: 17.0              ↓ flux: 28.6

Treatment II  | Upland | Rice season | Upland | Rice season |
```

Fig. 5 Two-year results of the effect of incorporation time of rice straw on mean CH_4 flux (mg CH_4 m^{-2} h^{-1}) over the rice growing period., incorporation time of same amount of rice straw into soil (Data cited from Xu et al., 2001).

It has been demonstrated that incorporating organic manure or crop residues, such as straw, during the aerobic (drained) season and allowing them to decompose in the soil under aerobic conditions for a certain period before the field is flooded significantly decreases the stimulation of CH_4 emissions during the subsequent flooded period (Shin et al., 1996; Wassmann et al., 2000). Our pot experiment supported this conclusion (Fig. 5). In this experiment, the same amount of rice straw was incorporated into the soil just after rice harvesting (Treatment I) and when the pot was prepared for rice transplanting (Treatment II). The mean CH_4 fluxes during the growing season were significantly higher in treatment II than in treatment I in consecutive 2-year experiments.

Except for permanently flooded rice fields, rice-based ecosystems in China have an aerobic period during which the fields are either planted with upland crops or left fallow under drained conditions. This period is longer than the growing season even in regions that produce two rice crops per year. Therefore, incorporating organic manure or crop residues into soils is most desirable when fields are under aerobic conditions. There are additional advantages to this approach. For instance, if rice straw or other organic manure is incorporated into the soil just before the field is flooded to permit rice transplanting, the organic matter makes it difficult for rice seedlings to stand. Rapid decomposition of the organic matter exhausts soil oxygen too quickly and promotes the development of excessively anaerobic conditions that can damage the development of rice roots. The application of organic manure and crop residues is also a demand for local environmental protection. Therefore, incorporation of organic manure and crop residues during the aerobic season should be encouraged.

References

Aulakh, M. S., Wassmann, R. and Rennenberg, H. (2002) Methane transport capacity of twenty-two rice cultivars from five major Asian rice-growing countries. *Agric. Ecosyst. Environ.* 91 (1-3): 59-71

Cai, Z. C., Xing, G. X., Yan, X. Y., Xu, H., Tsuruta, H., Yagi, K. and Minami, K. (1997) Methane and nitrous oxide emissions from rice paddy fields as affected by nitrogen fertilizers and water management. *Plant and Soil* 196 (1): 7-14

Cai, Z. C., Xing, G. X., Shen, G. Y., Xu, H., Yan, X.Y. and Tsuruta, H. (1999) Measurements of CH_4 and N_2O emissions from rice fields in Fengqiu, China. *Soil Sci. Plant Nutr.* 45: 1-13

Cai, Z. C., Tsuruta, H. and Minami, K. (2000) Methane emissions from rice fields in China: Measurements and influencing factors. *J. Geophys. Res.* 105 (D13): 17231-17242

Cai, Z. C., Tsuruta, H., Gao, M., Xu, H. and Wei, C. F. (2003) Options for mitigating methane emission from a permanently flooded rice field. *Global Changy Biol.* 9: 37-45

Huang, Y., Sass, R. L. and Fisher, F. M. (1998) Model estimates of methane emission from irrigated rice cultivation of China. *Global Change Biol.* 4 (8): 809-821

Inubushi, K., Hori, K., Matsumoto, S. and Wada, H. (1997) Anaerobic decomposition of organic carbon in paddy soil in relation to methane emission to the atmosphere *Water Sci. Technol.* 36 (6-7): 523-530

Kang, G. D., Cai, Z. C. and Fang, X. Z. (2002) Importance of water regime during the non-rice growing period in winter in regional variation of CH_4 emissions from rice fields during following rice growing period in China. *Nutrient Cycling in Agroecosystems* 64(1-2): 95-100

Kern, J. S., Gong, Z. T., Zhang, G. L., Zhuo, H. H. and Luo, G. B. (1997) Spatial analysis of methane emissions from paddy soils in China and the potential for emissions reduction. Nutrient Cycling in Agroecosystems 49: 181-195

Khalil, M. A. K., Rasmussen, R. A., Wang, M. X. and Ren, L. (1991). Methane emissions from rice fields in China. *Environ. Sci. Technol.* 25: 979-981

Lee, C. K. (1992) Paddy soil of China (in Chinese), Science Press, Beijing, China.

Li, J., Wang, M. X., Huang, Y. and Wang, Y. S. (2002) New estimates of methane emissions from Chinese rice paddies. *Nutrient Cycling in Agroecosystems* 64 (1-2): 33-42

Matthews, R. B., Wassmann, R., Knox, J. W. and Buendia, L. V. (2000) Using a crop/soil simulation model and GIS techniques to assess methane emissions from rice fields in Asia. IV. Upscaling to national levels. *Nutrient Cycling in Agroecosystems* 58 (1-3): 201-217

Office for the National Soil Survey. (1997) Data of Soil Survey of China. China Agricultural Press, China: 55-56

Sass, R. L., Fisher, F. M., Ding, A. and Huang, Y. (1999) Exchange of methane from rice fields: National, regional, and global budgets. *J. Geophys. Res. -Atmospheres* 104 (D21): 26943-26951

Shin, Y. K., Yun, S. H., Park, M. E. and Lee, B.L. (1996) Mitigation options for methane emission from rice fields in Korea. *AMBIO* 25 (4): 289-291

Ueki, A., Ono, K., Tsuchiya, A., Ueki, K. and Noike, K. (1997) Survival of methanogens in air-dried paddy field soil and their heat tolerance. *Water Sci. Technol.* 36: 517-522

Verburg, P. H. and Van der Gon H. A. C. D. (2001) Spatial and temporal dynamics of methane emissions from agricultural sources in China. *Global Change Biol.* 7 (1): 31-47

Wassmann, R., Papen, H. and Rennenberg, H. (1993) Methane emission from rice paddies and possible mitigation strategies. *Chemosphere* 26 (1−4): 201−217

Wassmann, R., Lantin, R. S., Neue, H. U., Buendia, L. V., Corton, T. M. and Lu, Y. (2000) Characterization of methane emissions from rice fields in Asia. III. Mitigation options and future research needs. *Nutrient Cycling in Agroecosystems* 58 (1−3): 23-36

Xie, D. T. (1988) Mechanisms of raising crop yield by natural free-tillage of rice fields (in Chinese). Ph. D. Thesis, Southwest China Agric. Univ., Chongqing, China.

Xu, H., Cai, Z. C. and Jia, Z. J. (2002) Effect of soil water contents in the non-rice growth season on CH_4 emission during the following rice-growing period. *Nutrient Cycling in Agroecosystems* 64 (1−2): 101−110

Xu, H., Cai, Z. C., Jia, Z. J. and Tsuruta, H. (2001) Effect of incorporation time of rice straw on CH_4 emissions from rice pots. *Agri. Environ. Protection* (in Chinese) 20: 26−28

Yagi, K., Minami, K. and Ogawa, Y. (1998) Effects of water percolation on methane emission from rice paddies: a lysimeter experiment. *Plant Soil* 198: 193−200

Yagi, K., Tsuruta, H. and Minami, K. (1997) Possible options for mitigating methane emission from rice cultivation *Nutrient Cycling in Agroecosystems* 49 (1−3): 213−220

Zheng, X. H., Wang, M. X., Wang, Y. S., Shen, R. X., Li, J., Heyer, J., Koegge, M., Papen, H., Jin, J. S. and Li, L. T., (2000) Mitigation options for methane, nitrous oxide and nitric oxide emissions from agricultural ecosystems. *Adv. Atmos. Sci.* 17 (1): 83−92

FACE : a window through which to observe the responses of agricultural ecosystems to future atmospheric conditions

Kazuhiko Kobayashi

National Institute for Agro-Environmental Sciences, 3-1-3 Kannondai, Tsukuba, Ibaraki 305-8604, Japan

Abstract

FACE (free-air CO_2 enrichment) lets researchers observe the responses of an intact ecosystem to elevated atmospheric CO_2 concentrations ($[CO_2]$) in the field without otherwise altering the canopy environment. From 1998 to 2000, we conducted a FACE experiment in rice in Japan, and found that both growth and yield were stimulated by elevated $[CO_2]$. Both changes depended on nitrogen availability: rice responded less to CO_2 enrichment at lower rates of nitrogen fertilization. We also found changes in soil-related activities, including higher methane emission rates and increased soil microbial biomass, in response to CO_2 enrichment. The results of FACE studies using other crops suggest that the rice responses were similar to those for C_3 grass species, but differed from the responses of leguminous species. Soil responses to elevated $[CO_2]$ varied among and within studies, warranting further studies before any generalizations can be made. Some ecosystem changes reported in FACE studies cannot be extrapolated to the future because the step-increase in $[CO_2]$ does not match the gradual $[CO_2]$ increase that occurs in the real world. However, the gap between experiments and reality could be filled by a process-based model of plant growth and biogeochemistry.

Keywords : CO_2, FACE, rice, soil microbes, biogeochemistry

1. Introduction to FACE

FACE (free-air CO_2 enrichment) provides a unique opportunity to observe the responses of intact terrestrial ecosystems to elevated atmospheric CO_2 concentrations ($[CO_2]$) without artificially altering the microclimate of the vegetation. Since its first conception as "free air stream CO_2 enrichment" by Herbert Z. Enoch in 1982 (Allen, 1992), we have seen considerable progress in the FACE method as an experimental technique, and increasing coverage of the ecosystems being studied (Norby et al., 2001).

Construction of the FACE system by scientists at the Brookhaven National Laboratory (BNL; Upton, New York, USA) proved for the first time that $[CO_2]$ can be raised reliably in the field without the use of any enclosures (Lewin et al., 1994). The invention of a mini-FACE system by Italian scientists has enabled small-scale FACE experiments by drastically shrinking the requirements for space and CO_2 supply (Miglietta et al., 2001a). We ourselves have proved that the injection of pure CO_2 at a high velocity could mix CO_2 with the ambient air as efficiently as in the BNL-designed FACE system, which is based on pre-dilution using blowers (Okada et al., 2001). This innovation is important because blowers have been found to change the microclimate of the vegetation and thereby alter plant developmental rates (Pinter et al., 2000). The injection of pure CO_2 has thus achieved the ideal sought by the inventor of FACE, and has now been used to observe the results of CO_2 enrichment in rice, soybean, maize, and a poplar plantation (Miglietta et al., 2001b), as well as in a natural forest (Haetenschwiler et al., 2002).

FACE is now being used in ca. 26 different ecosystems around the world (http://cdiac.esd.ornl.gov/programs/FACE/face.html). Among the FACE experiments, three are now being conducted in Asia: one in Japan, one in China, and one in India. It is noteworthy that these three experiments have the same crop species in common (rice, *Oryza sativa* L.), but differ in the crops being grown outside the rice-growing season. This is a unique situation that offers opportunities to compare plant and ecosystem responses to elevated $[CO_2]$ among soil types, climates, crop varieties, and agronomic practices on a common basis.

In this paper, I summarize the findings of the rice FACE experiment in Japan in the hope that the Japan − China collaboration will facilitate comparisons between the Japanese and Chinese results of plant and ecosystem responses to elevated $[CO_2]$.

2. Findings of the rice FACE experiment in Japan

We installed the newly designed FACE apparatus in paddy fields in Shizukuishi, near Morioka, Iwate Prefecture, in northern Japan (Fig. 1). The octagonal FACE ring consists of

Fig. 1 Overview of a FACE ring in a rice field in Shizukuishi, Japan

Table 1 Grain yield of the variety Akitakomachi in the Rice FACE experiment in Japan
See text for the nitrogen application rates.

Year	Unhulled fertile grain yield (dry matter metric ton ha^{-1}) (Values relative to the yield in Control plots in parentheses)					
	Low-N		Standard-N		High-N	
	Control	FACE	Control	FACE	Control	FACE
1998	5.14	5.73	5.88	6.81	6.47	7.84
	(1)	(1.11)	(1)	(1.16)	(1)	(1.21)
1999	6.86	7.14	7.67	8.70	8.36	9.31
	(1)	(1.04)	(1)	(1.13)	(1)	(1.11)
2000	6.37	6.78	7.23	8.28	7.85	8.87
	(1)	(1.07)	(1)	(1.15)	(1)	(1.13)

eight plastic tubes pierced by numerous tiny holes. Pure CO_2 is injected into the ambient air around the crop through the tiny holes in the tubes on the upwind sides of the structure, and mixes with the air while being transported downwind. The amount of CO_2 injected is computer-controlled so as to keep [CO_2] at the ring center close to the target level, which was chosen as 200 ppmV above ambient [CO_2] in the present experiment. The CO_2 enrichment was provided continuously from transplanting through harvesting of the rice plants for 3 years (from 1998 to 2000). The area within each FACE ring was split into subplots treated with different amounts of nitrogen (N) fertilizer so as to study the rice and soil responses to elevated [CO_2] under various levels of N availability. Control plots were also established, in which the field arrangements were identical to those of the FACE plots except for the omission of the CO_2-emission rings.

The grain yield of a Japanese rice cultivar ('Akitakomachi') was significantly increased by CO_2 enrichment in all 3 years and in all three N treatments (Table 1). At the standard N fertilization rate (80 kg N ha^{-1} in 1998, and 90 kg N ha^{-1} in 1999 and 2000), the grain yield in the FACE plots was 13% to 16% higher than in the control plots despite significant differences in actual yields observed in the controls due primarily to annual variations in climate. The response of grain yield to CO_2 enrichment was obviously and significantly influenced by the N fertilization rate. The magnitude of the yield response to low levels of N fertilization (40 kg N ha^{-1} in 1998, 1999 and 2000) was only about a half the response in the treatments with standard N or a high level of N fertilization (120 kg N ha^{-1} in 1998, and 150 kg N ha^{-1} in 1999 and 2000).

Analysis of the yield components showed that the increased number of spikelets per unit of ground area was the major determinant of the response to elevated [CO_2] as well as to N fertilization (Kim et al., 2003a). This is particularly interesting because the number of grains is determined around the panicle initiation stage (PI), which represents only the midpoint of the duration of growth and occurs well before the accumulation of carbohydrates begins in the rice grains.

Further analysis showed that N accumulation through the PI stage is the key determinant of the number of grains, and that the same relationship holds across [CO_2] levels (Fig. 2). Up to the PI stage, CO_2 enrichment increased N accumulation, and thereby increased the number of grains in the standard and high-N fertilization treatments. In the low-N treatment, N uptake through the PI stage was unchanged by CO_2 enrichment, and the number of grains increased much less than at higher N levels.

At the grain-filling stage, the larger number of grains present under elevated [CO_2] demanded more N for grain growth than under ambient [CO_2] in the standard and high-N

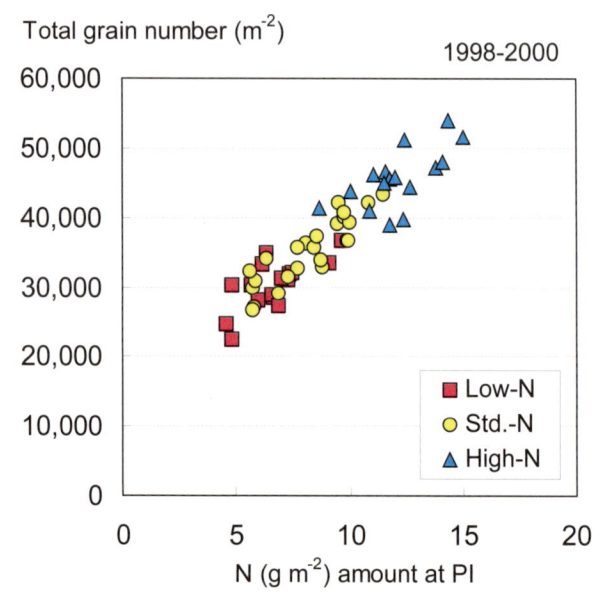

Fig. 2 Relationship between number of fertile grains at harvest and nitrogen accumulated through to PI

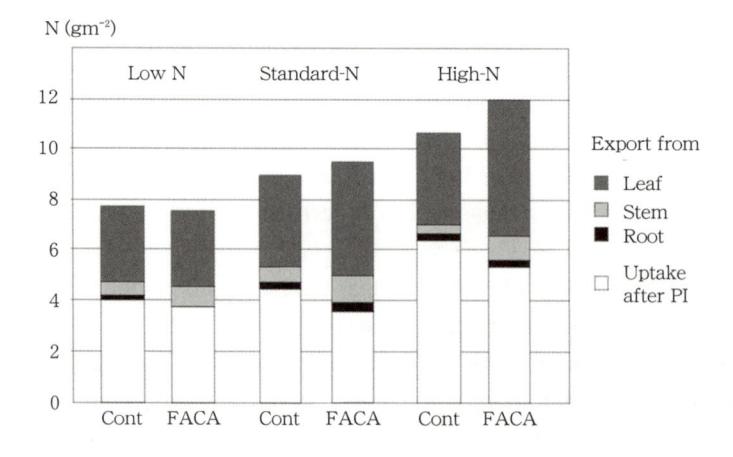

Fig. 3 Apparent sources of nitrogen in grains at maturity in the Rice FACE experiment in 1999

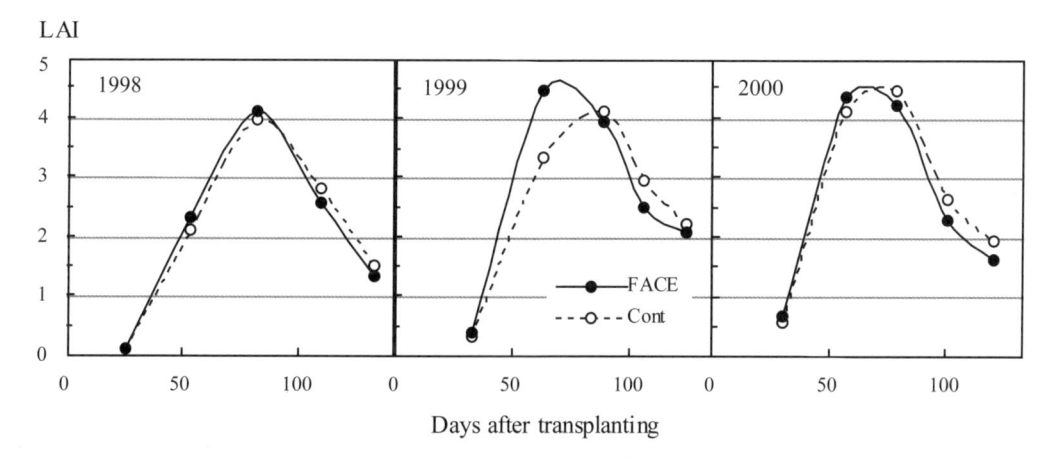

Fig. 4 Changes in green leaf area index (LAI) under elevated (FACE) and ambient (Cont) CO_2 concentrations in the standard-N fertilization in the Rice FACE experiment

treatments. This was not the case in the low-N plots. The higher demand for N by the grains under elevated $[CO_2]$ was met by increased retranslocation of N from the leaves and stems, whereas N uptake after the PI stage was lower under elevated $[CO_2]$ (Fig. 3).

A consequence of the larger export of N from the leaves to the grains is quicker senescence of the leaves and, hence, an earlier loss of photosynthesizing leaf area in the CO_2-enriched plants (Fig. 4). The green leaf area index (GLAI) under elevated $[CO_2]$ peaked around the PI stage in 1999 and 2000 (about 60 days after transplanting), whereas GLAI at ambient $[CO_2]$ was highest at the flowering stage (about 80 days after transplanting). The export of N from the leaves to the developing panicles might have also contributed to a decline in photosynthetic capability of the flag leaves that has been observed under CO_2 enrichment (Seneweera et al., 2002). These declines in leaf area and photosynthetic capability in the CO_2-enriched plants could account for the lack of an effect of FACE on net biomass increase during the period between PI and maturity.

The question is, then, how grain yields increased without a concomitant increase in total biomass accumulation during the grain-filling period. The higher rate of leaf loss under the CO_2 enrichment could, at least partially, have cancelled out the larger increase in grain mass, resulting in the small difference in net biomass increase during the grain-filling period between $[CO_2]$ levels.

In addition to these changes in rice growth and yield, we found changes in soil-related processes under elevated $[CO_2]$. Methane (CH_4) emission from the rice paddies was determined by means of the closed-chamber method every 2 weeks in 1999 and 2000. The results showed that the seasonal integral CH_4 emission increased by 38% (1999) and 51% (2000) due to CO_2 enrichment (Inubushi et al., 2003). The increase in CH_4 emission caused by the $[CO_2]$ increase warrants further study so we can understand the mechanisms involved and better quantify the increase, since the CO_2-CH_4 interaction could exacerbate the greenhouse effect caused by the individual gases.

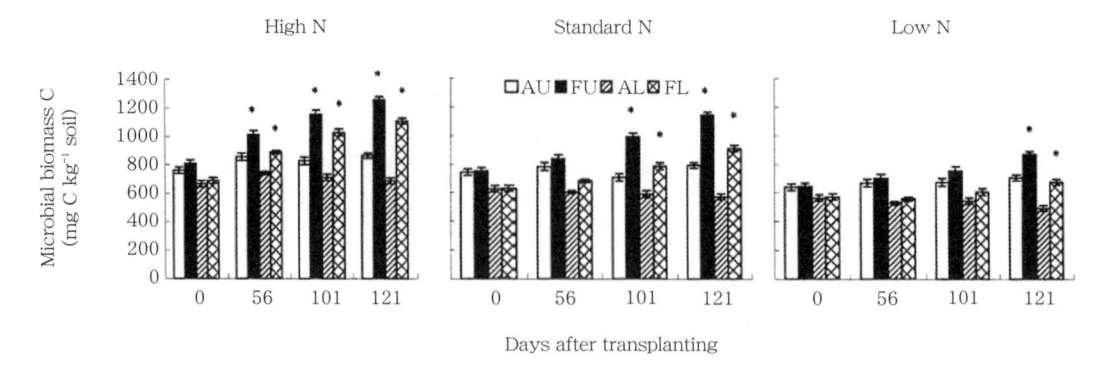

Fig. 5 Soil microbial biomass carbon in upper: U (0~1 cm) and lower : L (1~10 cm) soil layers of FACE (F) and control (A) plots during the growing season of 2000. Statistical significance between the FACE and ambient CO_2 treatments are indicated by *

Soil microbial processes also responded to elevated [CO_2], but depended on N availability to some extent. As shown in Figure 5, soil microbial biomass carbon (SMC) was increased by CO_2 enrichment at the high-N fertilization level from the PI stage on, but at the low-N level, this increase was observed only at harvest (Hoque et al., 2001). Soil microbial biomass nitrogen (SMN) responded to CO_2 enrichment differently from SMC (data not shown). SMN was increased by CO_2 enrichment by the time of harvest, but the extent tended to be greater in the low-N treatment; however, the CO_2-N interaction was not statistically significant (Hoque et al., 2002).

The discrepancy between the SMC and SMN responses to elevated [CO_2] may indicate changes in the composition of the microbial species as a result of changes in C and N availability in the soil. The larger root biomass of the rice plants under elevated [CO_2] (Kim et al., 2003b) should have increased the C flux into the soil as a result of root turnover. The increase in the biomass of algae and floating weeds in the water layer (Koizumi et al., 2001) may also have increased C and N availability at the soil surface.

3. Synthesis of FACE findings across crop species

How do our findings compare with the results of other FACE studies? Kimball et al. (2002) summarized the data from several publications on FACE experiments that studied agricultural crops, and reviewed the plant and soil responses to [CO_2] scaled to a 200 ppmV elevation above the current ambient level. The crop species included in their review were cotton, wheat, sorghum, rice, ryegrass, white clover, lucerne, grape, and potato. Although the coverage of species was by no means sufficient for drawing broad conclusions, the species represented a range of different types of crop species: C_3 grasses (wheat, rice, and ryegrass), C_4 grass (sorghum), legumes (clover and lucerne), C_3 broadleaf forb (potato), and C_3 woody perennials (cotton and grape).

The different types of crop exhibited different responses to elevated [CO_2]. In terms of the aboveground plant biomass, the C_3 grasses under deficient N supply showed a smaller CO_2

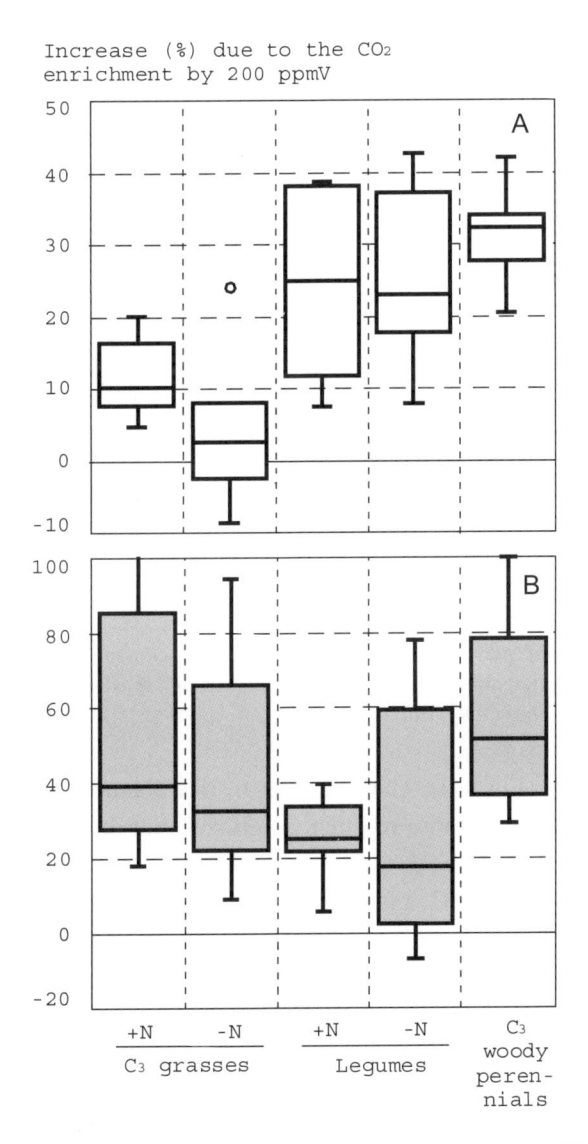

Fig. 6 Responses of the above-ground (A) and below-ground (B) biomass to CO_2 enrichment with affluent N supply (+N) or deficient N supply (-N) in C_3 grasses, legumes, and C_3 woody perennials

response than those under an ample N supply (Fig. 6A). The legumes, by comparison, showed a larger response to elevated [CO_2] than did the C_3 grasses, but the response under low N availability was not significantly different from the response under ample N supply (Fig. 6A). Although the greater response in the legumes than in the C_3 grasses could be ascribed to symbiotic N_2 fixation in the legumes, the C_3 woody perennials showed a response comparable to that of the legumes, even though these plants lacked N_2 fixation (Fig. 6A). Root biomass in the C_3 grasses clearly showed a greater response to elevated CO_2 than did aboveground biomass. In contrast, the root response in the legumes did not differ from those

```
Increase (%) due to the CO₂
enrichment by 200 ppmV
```

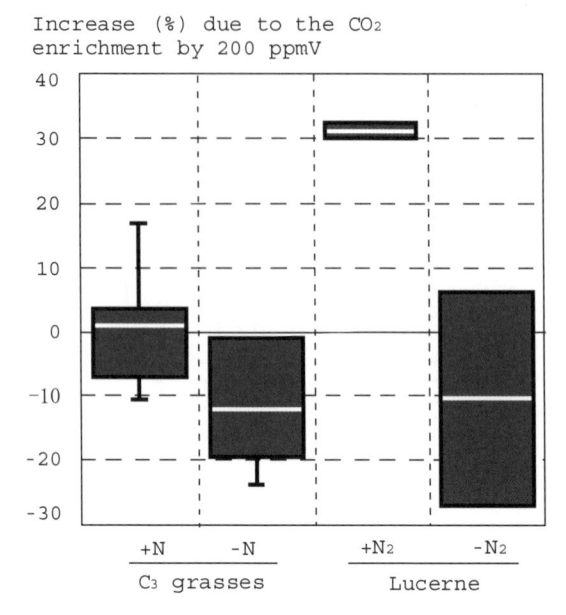

Fig. 7 Response of plant N accumulation to CO_2 enrichment with affluent N supply (+N) or deficient N supply (−N) in C_3 grasses, and N_2-fixing lines (+N_2) or non-N_2 fixing lines (−N_2) of lucerne

of aboveground biomass (Fig. 6B). The responses of rice plants to elevated CO_2 (Kim et al., 2003b) are thus comparable to those of other C_3 grasses, but different from the responses of leguminous species.

The discrepancy between the C_3 grasses and the legumes in their CO_2 response was also evident in plant N uptake. In the C_3 grasses, N accumulation per unit of ground area was unchanged or was slightly reduced by CO_2 enrichment (Fig. 7). In lucerne, a legume, the N_2-fixing lines showed increased N accumulation in response to CO_2 enrichment, whereas lines that lacked the ability to fix N_2 showed a response similar to that of the C_3 grasses (Fig. 7). The non-N_2-fixing lucerne also showed a plant biomass response similar to that of the C_3 grasses (Kimball et al., 2002). It thus appears that the ability to fix N_2 explains the difference between the legumes and the C_3 grasses in their responses to elevated CO_2. The lack of response of N accumulation to elevated [CO_2] may be the cause of the smaller increase in plant biomass in the C_3 grass species than in the legumes.

Soil responses to elevated [CO_2] were more variable than the plant responses, and general patterns could barely be discerned in Kimball et al.'s (2002) review. This is partly because of the paucity of observations focusing on soil processes in previous FACE studies. Like the different plant responses to elevated [CO_2], different soil-crop combinations may exhibit different responses to elevated [CO_2], and the large variability in the FACE results may simply have arisen because of such variability. In spite of the large variability, however, Kimball et al. (2002) found a tendency for soil organic C concentration to be higher under elevated [CO_2] across all the FACE studies.

Soil microbial activities were even more variable in their response to CO_2 enrichment. Only a FACE study with ryegrass and white clover observed soil microbial C, which was found to have increased owing to CO_2 enrichment, although the effect was not statistically significant (Kimball et al., 2002). Zak et al. (2000) have reviewed a larger number of soil microbial studies with CO_2 enrichment using FACE and growth chambers, and also found high variability across studies in microbial biomass responses. It is not therefore possible to generalize the rice FACE findings on soil microbial responses to elevated $[CO_2]$. Further studies of microbial processes under FACE are thus warranted because of the critical roles that soil microbes play in linking plants and soils via their C and N metabolisms. Without understanding why there is such a large degree of variability in the responses of soil processes to elevated $[CO_2]$, we will not be able to predict soil C and N cycling under higher $[CO_2]$ in the future (Zak et al., 2000).

4. Process-based modeling based on the rice FACE findings

The synthesis of FACE results across several crop species discussed in the previous section has revealed variability among the crop types in the plant responses to elevated $[CO_2]$. It is thus important to determine the causes of variability such as the smaller CO_2 response under lower N availability or the higher CO_2 response in the leguminous species. To address these questions, it would be necessary to assemble individual FACE findings to provide a more complete understanding of the entire ecosystem's responses to elevated $[CO_2]$.

As a vehicle for better understanding of rice ecosystem responses to elevated $[CO_2]$, a biogeochemistry model called DNDC (DeNitrification and DeComposition) shows considerable promise (Li, 2000). This model has been modified to accommodate plant physiological responses to elevated $[CO_2]$, and is being tested against the rice FACE observations in Japan. Once the model has been successfully verified, it will be tested against the FACE results in China as well as the FACE observations to be conducted in Japan in 2003 and 2004. Collaboration with Indian scientists is also envisioned so that the model can be tested against the FACE results in India. It is hoped that the many pieces of evidence from the FACE experiments can be assimilated into the modified model. The model can then be used both to predict the responses of future agricultural ecosystems and to appraise the efficacy of agronomic adaptations to future $[CO_2]$.

Biogeochemistry is a crucial component of any model designed to simulate the ecosystem responses to increasing $[CO_2]$. Although the physiological responses of annual plants to higher $[CO_2]$ take place on time scales ranging from seconds to days, some soil processes may respond very slowly to higher $[CO_2]$, on the scale of decades to centuries. In the long run, however, these slow changes in soil properties may be crucial in determining the changes in plant productivity. It must also be noted that the increase in $[CO_2]$ in the real world is gradual, unlike the step-increase observed in FACE experiments. We cannot therefore directly extrapolate what we see in FACE experiments into future ecosystems (Luo and Reynolds, 1999), but we can use the findings to improve process-based models that can then provide a better view of the future.

5. Conclusions

Atmospheric [CO_2] increases will affect the carbon and nitrogen metabolisms of both plants and soil organisms. These changes will also change the fluxes of trace gases and possibly the composition of the soil microbial community. The changes in fluxes of trace gases will also alter air chemistry, with effects that are more difficult to predict because these gases have been less well studied. The potential changes in the soil microbial community may alter soil characteristics and crop productivity in both the short and the long run; simultaneous changes in biogeochemical processes may have additional complicating effects. The results of the FACE experiments cannot be simply extrapolated to predict the future situation, since the step-increase in [CO_2] used in the experiments differs from the gradual increases observed in nature, but will nonetheless be valuable in creating models that are better suited to predicting ecosystem responses to increasing atmospheric [CO_2].

Acknowledgments

The rice FACE project was supported by CREST (Core Research for Evolutional Science and Technology) of the Japan Science and Technology Corporation (JST), under contracts between JST and the National Institute for Agro-Environmental Sciences, and between CREST and the Tohoku National Agricultural Experiment Station.

References

Allen, L. H. Jr. (1992) Free-air CO_2 enrichment field experiments: An historical overview. *Crit. Rev. Plant Sci.* 11: 121-134

Haetenschwiler, S., Handa, I. T., Egli, L., Asshoff, R., Ammann, W. and Koerner, C. (2002) Atmospheric CO_2 enrichment of alpine treeline conifers. *New Phytol.* 156: 363-375

Hoque, Md. M., Inubushi, K., Miura, S., Kobayashi, K., Kim, H. Y., Okada, M. and Yabashi, S. (2001) Biological dinitrogen fixation and soil microbial biomass carbon as influenced by free-air carbon dioxide enrichment (FACE) at three levels of nitrogen fertilization in a paddy field. *Biol. Fertil. Soils* 34: 453-459

Hoque, Md. M., Inubushi, K., Miura, S., Kobayashi, K., Kim, H. Y., Okada, M. and Yabashi, S. (2002) Nitrogen dynamics in paddy field as influenced by free-air CO_2 enrichment (FACE) at three levels of nitrogen fertilization. *Nutrient Cycling in Agroecosystems* 63: 301-308

Inubushi, K., Cheng, W., Aonuma, S., Hoque, Md. M., Kobayashi, K., Miura, S., Kim, H. Y. and Okada, M. (2003) Effects of free-air CO_2 enrichment (FACE) on CH_4 emission from a rice paddy field. *Global Change Biol.* 9: 1458-1464

Kim, H. Y., Lieffering, M., Kobayashi, K., Okada, M., Mitchell, M. and Gumpertz, M. (2003a) Effects of free-air CO_2 enrichment and nitrogen supply on the yield of temperate paddy rice crops. *Field Crops Res.* 83: 261-270

Kim, H. Y., Lieffering, M., Kobayashi, K., Okada, M. and Miura, S. (2003b) Seasonal changes in the effects of elevated CO_2 on rice at three levels of nitrogen supply: a free air CO_2 enrichment (FACE) experiment. *Global Change Biol.* 9: 826-837

Kimball, B. A., Kobayashi, K. and Bindi, M. (2002) Responses of agricultural crops to free-air CO_2 enrichment. *Advances in Agron.* 77: 293–368

Koizumi, H., Kibe, T., Mariko, S., Ohtsuka, S., Nakadai, T., Mo, W., Toda, H., Nishimura, S. and Kobayashi, K. (2001) Effect of free-air CO_2 enrichment (FACE) on CO_2 exchange at the flood-water surface in a rice paddy field. *New Phytol.* 150: 231–239

Lewin, K. F., Hendrey, G. R., Nagy, J. and LaMorte, R. L. (1994) Design and application of a free-air carbon dioxide enrichment facility. *Agric. For. Meteorol.* 70: 15–29

Li, C. (2000) Modeling trace gas emissions from agricultural ecosystem. *Nutrient Cycling in Agroecosystems* 58: 259–276

Luo, Y. and Reynolds, J. F. (1999) Validity of extrapolating field CO_2 experiments to predict carbon sequestration in natural ecosystems. *Ecology* 80: 1568–1583

Miglietta, F., Hoosbeek, M. R., Foot, J., Gigon, F., Hassinen, A., Heijmans, M., Peressotti, A., Saarinen, T., van Breemen, N. and Wallen, B. (2001a) Spatial and temporal performance of the miniFACE (Free Air CO_2 Enrichment) system on bog ecosystems in northern and central Europe. *Environ. Monitoring Assess.* 66: 107–127

Miglietta, F., Peressotti, A., Vaccari, F. P., Zaldei, A., deAngelis, P. and Scarascia-Mugnozza, G. (2001b) Free-air CO_2 enrichment (FACE) of a poplar plantation: the POPFACE fumigation system. *New Phytol.* 150: 465–476

Norby, R. J., Kobayashi, K. and Kimball, B. A. (2001) Rising CO_2-future ecosystems. *New Phytol.* 150: 215–221

Okada, M., Lieffering, M., Nakamura, H., Yoshimoto, M., Kim, H. Y. and Kobayashi, K. (2000) Free-air CO_2 enrichment (FACE) with pure CO_2 injection: system description. *New Phytol.* 150: 251–260

Pinter, P. J. Jr., Kimball, B. A., Wall, G. W., LaMorte, R. L., Hunsaker, D. J., Adamsen, F.J., Frumau, K. F. A., Vugts, H. F., Hendrey, G. R., Lewin, K. F., Nagy, J., Johnson, H. B., Wechsung, F., Leavitt, S. W., Thompson, T. L., Mattias, A. D. and Brooks, T. J. (2000) Free-air CO_2 enrichment (FACE): blower effects on wheat canopy microclimate and plant development. *Agric. For. Meteorol.* 103: 319–333

Seneweera, S. P., Conroy, J. P., Ishimaru, K., Ghannoum, O., Okada, M., Lieffering, M., Kim, H. Y. and Kobayashi, K. (2002) Changes in source-sink relations during development influence photosynthetic acclimation of rice to free air CO_2 enrichment (FACE). *Functional Plant Biol.* 29: 945–953

Zak, D. R., Pregitzer, K. S., King, J. S. and Holmes, W. E. (2000) Elevated atmospheric CO_2, fine roots and the response of soil microorganisms: a review and hypothesis. *New Phytol.* 147: 201–222

Estimating carbon sequestration in Japanese arable soils using the Rothamsted carbon model

Yasuhito Shirato and Masayuki Yokozawa

National Institute for Agro-Environmental Sciences, 3-1-3, Kannondai
Tsukuba, Ibaraki 305-8604, Japan

Abstract

The Rothamsted Carbon Model (RothC) was tested using long-term experimental datasets from Japan. RothC simulated changes in soil organic carbon (SOC) well for most major soil types, but underestimated the measured SOC for Andosols. We modified the model to suit Andosols by changing the decomposition rate for the humus pool because Al-humus complexes in Andosols stabilize the humus against decomposition. We also set the inert organic matter pool to zero. We obtained greatly improved fits between modeled and measured SOC using the modified model. We then applied the model to all of Japan using 1-km-resolution grid datasets based on land use, weather, and soil types. We calculated the amount of C input to soils by assuming that SOC reaches an equilibrium. We then simulated changes in SOC after application of farmyard manure at 0.5 or 1.0 t C ha^{-1} yr^{-1} for 50 yr from the present. After 50 yr, more than 5.6 or 11.1 Mt carbon, respectively, had been stored. This amount is large compared to the desired reduction of CO_2 emissions under the Kyoto protocol. We therefore suggest that arable soils could be a large C sink under appropriate management.

Key words : Andosol, Japan, long-term experiment, modeling, soil organic matter

1. Introduction

The Kyoto Protocol allows carbon emissions to be offset by demonstrable removals of carbon from the atmosphere. Activities that have been shown to reduce atmospheric CO_2 levels include improved management of agricultural soils as well as afforestation and reforestation. We must therefore provide good methods to evaluate changes in the stores of soil organic carbon (SOC) caused by management changes, including the management of arable soils.

Soils can serve as both sources of and sinks for CO_2. Because some changes in SOC content or composition occur slowly, over the course of decades, long-term experiments are useful for measuring these changes. Because it is impossible to carry out long-term experiments for all soil types with many management scenarios, models of the turnover of soil organic matter (SOM) are the only practical means of estimating SOC changes in places other than the sites of long-term experiments.

Many SOM turnover models have been published. We selected the Rothamsted Carbon Model (RothC) as a tool for estimating C sequestration in Japanese arable soils because the few inputs required by the model are easily obtainable. This means that the model can be applied over wide areas, such as the whole of a country, with a reasonable expenditure of resources and time.

The current version of RothC (RothC-26.3) is a descendant of earlier versions developed by Jenkinson and Rayner (1977). It has been used to simulate changes in SOC for a variety of soil types and for a range of land uses, including arable land, grassland, and forest (Coleman and Jenkinson, 1996; Jenkinson et al., 1999a, 1999b). It has also been used at the regional scale (Falloon et al., 1998b, 2002; Falloon and Smith, 2002) and at the global scale (Jenkinson et al., 1991; Polglase and Wang, 1992). However, it has not yet been rigorously tested in Asian countries, including Japan.

The goals of this study were to test predictions by RothC against measured values of Japanese arable soils at the scale of individual plots, then estimate C sequestration at the scale of all fields of upland crops in Japan.

2. Materials and Methods

2.1 Model validation at the plot scale using long-term experimental datasets

We selected 10 long-term experiments to validate the model: four on Andosols, two on Brown Lowland Soils, two on Yellow Soils, one on Brown Forest Soil, and one on Gray Lowland Soil. In modeling each set of experimental data, it was first necessary to run the model to produce an initial SOC content. The amount of C input from plant residues was calculated by running the model backwards assuming that SOC was at equilibrium when the experiment started. A ratio of decomposable plant material (DPM) to resistant plant material (RPM) was defined on the basis of the original vegetation. The pool of inert organic matter (IOM) was calculated with the equation of Falloon et al. (1998a). Weather data were obtained from the Japan Meteorological Agency (1996). The potential evapotranspiration described by Thornthwaite (1948) was used to simulate open-pan evaporation. Soils were assumed be covered with vegetation throughout the year.

Once the initial SOC was established, we ran the model forward using the C inputs and soil cover information defined for each treatment. The amount of C input from plant residue was estimated from the proportion of dry matter production for each part of the plant, using the data of Ogawa et al. (1988). Plant input of C was added to the soils only in the month in which harvesting occurred. A DPM/RPM ratio of 1.44, which was recommended by the model developer (e.g. Coleman and Jenkinson, 1996), was used for all kinds of plant residue. The same manure quality (DPM = 49%, RPM = 49%, humus = 2%) was used for all manures. Soil cover information was based on the dates of sowing and harvesting.

2.2 Modifying RothC to suit Andosols

The high SOC level in Andosols is caused by the presence of active Al or Fe derived from volcanic ash; these metals form a stable, decomposition-resistant complex with humus (e.g., Shoji et al., 1993, pp. 162–163; Wada and Higashi, 1976). We therefore decided to modify RothC to account for this unique property of the humus in Andosols. Parfitt and Childs (1988) summarized the amounts of active Al and Fe in a soil as follows: pyrophosphate-extractable Al (Alp) corresponds to the amount of Al in humus complexes; Al and Si in allophane and imogolite are estimated from acid-oxalate-extractable Al (Alo), minus Alp and acid-oxalate-extractable Si (Sio); and ferrihydrite can be estimated from acid-oxalate-extractable Fe (Feo).

We used 32 data sets from the Andisol TU Database 1992 (Shoji et al., 1993, p. 263) to modify RothC. We used a representative C input value for each type of vegetation: 2.0, 3.8, 4.0, 3.3, and 3.0 t C ha^{-1}y^{-1} for woodland, grassland, *Miscanthus sinensis*, *Sasa*, and shrubs, respectively. We used DPM/RPM ratios of 1.44 for grassland, 0.25 for woodland, and 0.67 for the other vegetation types. Soils were assumed be covered with vegetation throughout the year. We set the IOM value to zero for all sites since the soil initially contained no carbon when it was formed from fresh volcanic ash.

We ran RothC-26.3 to produce estimates of SOC at equilibrium, and used a conversion factor H(f) to change the decomposition rate of the humus pool to fit the data calculated for each of the 32 data sets. We then performed regressions with H(f) being the response variable and Alp, Alo, Feo, Sio, and their combinations used as explanatory variables. The best fit was found using equation 1:

$$H(f) = 1.20 + 2.50 Alp \qquad (r^2 = 0.52) \qquad (1)$$

We tested the modified model, incorporating this correction factor, against data from four long-term experiments on Andosols using the same procedure as in section 2.1.

2.3 Estimating carbon sequestration by applying RothC to all Japanese upland crops

To apply the model for all Japanese upland crops, we constructed 1-km-resolution grid datasets on land use, weather, and soil types for all of Japan. We extracted grids in which the main land use was upland crops. The estimated total SOC for the extracted grids was about 109 Mt. We calculated the amount of C input from plant residues by running the model backwards and assuming that SOC was at equilibrium. The modified RothC model was used for Andosols; the original model was used for all other soils. We then simulated the changes in SOC after the application of farmyard manure (FYM) at 0.5 or 1.0 t C ha^{-1} every year from the present until the end of a 50-yr simulation period. In this simulation, SOC increased rapidly at first, then continued increasing slowly, but did not reach an equilibrium state during the 50-yr period.

3. Results and Discussion

3.1 Model validation

Figure 1 compares the modeled vs. measured changes in SOC on non-volcanic soils. The

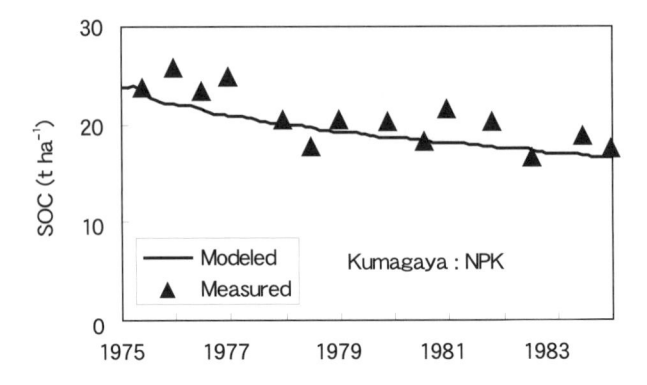

Fig. 1 Modeled vs.measured SOC on non-volcanic soils. Example from Gray lowland soils in Kumagaya, Saitama prefecture

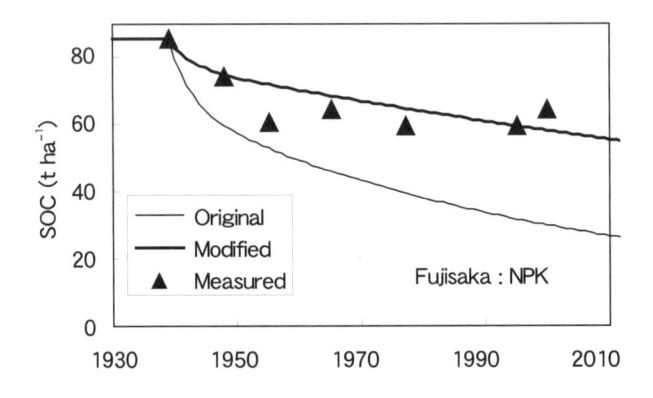

Fig. 2 Modeled vs.measured SOC on Andosols by original RothC and modified model. Example from Fujisaka, Aomori prefecture

SOC values predicted by the model were close to the measured values. These results indicate that the model was able to simulate the time course of changes in SOC well.

Figure 2 compares the modeled vs. measured time course of changes in SOC for Andosols. The SOC values obtained by the modified model were quite close to the measured values in almost all treatments at all four study sites, whereas those obtained by the original model were much lower than the measured values. These results indicate that the modified model was able to simulate changes in SOC better than the original model in almost all plots.

We were able to obtain good fits between the modeled and measured changes in SOC over time in almost all treatments at 10 sites across Japan using the modified model for Andosols and the original model for other soils. The 10 experimental sites were located from north to south across Japan, with mean annual air temperatures ranging from 5.2 to 16.4 °C, mean annual precipitation ranging from 786 to 2542 mm, and soil clay content ranging from 11.8%

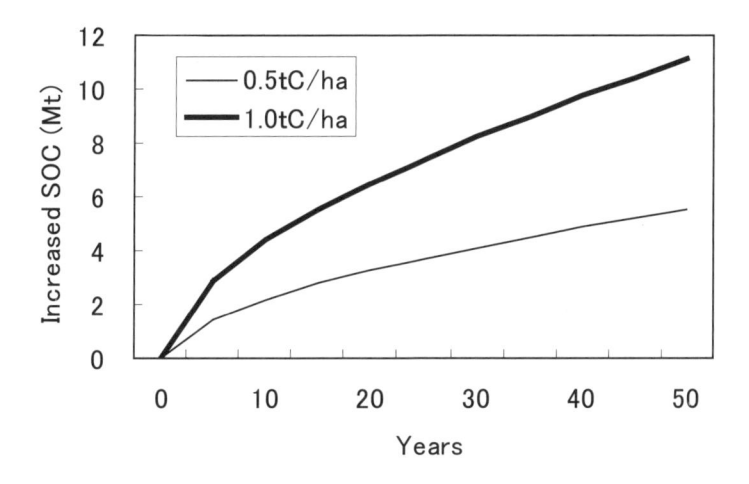

Fig. 3 Increased SOC in all Japanese upland crop fields by annual FYM application scenarios. Initial SOC was 109 Mt

to 38.0%. Moreover, each of the 10 sites had been subjected to different treatments (e.g., different C input from different crops with or without FYM). Thus, the model was tested and found to be applicable under various combinations of climatic conditions, soil textures, and land management scenarios.

3.2 Estimating carbon sequestration for all upland field crops in Japan

The simulated SOC increased over time, fastest initially, then gradually decreasing. However, SOC did not reach equilibrium during the 50-year simulation period. More than 5.6 or 11.1 Mt carbon was stored in all fields of Japanese upland crops after 50 yr at a FYM application rate of 0.5 or 1.0 t C ha^{-1}, respectively (Fig 3). This increase in SOC corresponds to 5% to 10% of the total SOC in all upland crops.

Under the Kyoto protocol, Japan is committed to reducing CO_2 emissions to 94% of the baseline (1990) levels during the first commitment period (2008-2012). This corresponds to a reduction in CO_2 emissions of 20 Mt C. The amount of C stored in soils through increased SOC under the scenarios in our study could contribute a significant portion of this total. The FYM input of 0.5 to 1.0 t C ha^{-1} is possible to realize in practice. We therefore suggest that arable soils can serve as a large C sink. Future studies must estimate the changes of SOC in paddy soils, since these soils are an important feature of many Asian countries.

Acknowledgments

We thank Prof. David Jenkinson and Dr. Kevin Coleman (IACR-Rothamsted) for advice on RothC; Dr. Ichiro Taniyama (NIAES) for useful discussion on the paper; Drs. Harushi Kikuchi (Fujisaka branch of the Aomori Agricultural Experiment Station), Hiroyuki Siga and Kazuo Konno (Kitami Agricultural Experiment Station), Toshifumi Murakami and Seishi

Yoshida (Nagano Chu-sin Agricultural Experiment Station), and Hidemi Wakikado and Hiroharu Furue (Kagoshima Agricultural Experiment Station) for assisting with soil sampling and data collection.

This work was supported financially by the Global Environment Research Fund of the Ministry of Environment, Japan (K-1 Carbon Sink Function of Terrestrial Ecosystems) from 1999 to 2001.

References

Coleman, K. and Jenkinson, D. S. (1996) RothC-26.3 - A model for the turnover of carbon in soil. In: Evaluation of Soil Organic Matter Models, Using Existing Long-Term Datasets. (ed. by Powlson, D. S., Smith, P. and Smith, J. U.) Springer, Berlin. pp. 237-246.

Falloon, P. and Smith, P. (2002) Simulating SOC changes in long-term experiments with RothC and CENTURY: model evaluation for a regional scale application. *Soil Use Manage.* 18: 101-111

Falloon, P., Smith, P., Coleman, K. and Marshall, S. (1998a) Estimating the size of the inert organic matter pool from total soil organic carbon content for use in the Rothamsted carbon model. *Soil Biol. Biochem.* 30: 1207-1211

Falloon, P. D., Smith, P., Smith, J. U., Szabo, J., Coleman, K. and Marshall, S. (1998b) Regional estimates of carbon sequestration potential: linking the Rothamsted Carbon Model to GIS databases. *Biol. Fertil. Soils* 27: 236-241.

Falloon, P. D., Smith, P., Szabó, J., Laszló Pásztor, Smith, J. U. and Coleman, K. (2002) Comparing estimates of regional carbon sequestration potential using geographical information systems, dynamic soil organic matter models, and simple relationships. In: Agricultural Practices and Policies for Carbon Sequestration in Soil. (ed. By Kimble, J. M., Lal, R., and Follett, R. F.). Lewis Publishers, Boca Raton. pp. 141-154

Japan Meteorological Agency (1996) Normal values on surface meteorological observation. Japan Meteorological Agency, Tokyo, Japan. (CD-ROM)

Jenkinson, D. S. and Rayner, J. H. (1977) The turnover of soil organic matter in some of the Rothamsted classical experiments. *Soil Sci.* 123: 298-305

Jenkinson, D. S., Adams, D. E. and Wild, A. (1991) Model estimates of CO_2 emissions from soil in response to global warming. *Nature* 351: 304-306

Jenkinson, D. S., Harris, H. C., Ryan, J., McNeill, A. M., Pilbeam, C. J. and Coleman, K. (1999a.) Organic matter turnover in a calcareous clay soil from Syria under a two-course cereal rotation. *Soil Biol. Biochem.* 31: 687-693

Jenkinson, D. S., Meredith, J., Kinyamario, J. I., Warren, G. P., Wong, M. T. F., Harkness, D. D., Bol, R. and Coleman, K. (1999b) Estimating net primary production from measurements made on soil organic matter. *Ecology* 80: 2762-2773

Ogawa, K., Takeuchi, Y. and Katayama, M. (1988) Biomass production and the amounts of absorbed inorganic elements by crops in arable lands in Hokkaido, and its evaluation. *Res. Bull. Hokkaido Nat. Agric. Exp. Stn.* 149: 57-91

Parfitt, R. L. and Childs, C. W. (1988) Estimation of forms of Fe and Al: a review, and analysis of contrasting soils by dissolution and Moessbauer methods. *Austral. J. Soil Res.* 26: 121?144

Polglase, P. J. and Wang, Y. P. (1992) Potential CO_2 induced carbon storage by the terrestrial biosphere. *Austral. J. Bot.* 40: 641–656

Shoji, S., Nanzyo, M. and Dahlgren R. A. (1993) *Volcanic Ash Soils*. Elsevier, Amsterdam. 288 pp.

Thornthwaite, C. W. (1948) An approach toward a rational classification of climate. *Geogr. Rev.* 38: 55–94

Wada, K. and Higashi, T. (1976) The categories of aluminium- and iron-humus complexes in Ando soils determined by selective dissolution. *J. Soil Sci.* 27: 357–368

Studies on the coordination of the soil phosphorus supply with rice photosynthesis under different nutrient treatments

Changming Yang[a], Linzhang Yang[a] and Toshiaki Tadano[b]

[a] Institute of Soil Science, Chinese Academy of Sciences, Nanjing 210008, China
[b] Faculty of Agriculture, Hokkaido University, Sapporo 060-8589, Japan

Abstract

A 4-yr field experiment was conducted from 1999 to 2002 to study the coordination of the soil phosphorus (P) supply with photosynthesis by comparing the dynamics of soil P and the photosynthetic rate of rice plants under different nutrient treatments. We also explored possible mechanisms for this coordination by contrasting and analyzing the effectiveness of nutrient treatments on P absorption, root morphology, and acid phosphatase activity (APA) at the root surface. The combined application of chemical fertilizers and organic manure markedly increased soil P during the middle and later growth stages, and the extent to which soil P coordinated with photosynthesis was greatest after booting of the rice. The integration of inorganic and organic nutrient sources, particularly farm yard manure, also increased root-surface APA and improved root morphology during the middle and later growth stages, ultimately resulting in a greater proportion of P uptake by rice plants and a higher photosynthetic rate at the grain-filling stage than was the case with exclusively chemical fertilizers.

Key words : acid phosphatase activity, farmyard manure, soil P, paddy soil, photosynthetic rate

1. Introduction

The intensification of livestock and crop production has led to a large increase in the production of animal waste and straw in many regions of the world, including China. Recycling of animal manure and crop straw to agricultural land can supply valuable quantities of plant nutrients and soil organic matter, and can help meet crop nutrient requirements while maintaining soil fertility (Hadas et al., 1990). It has been well documented that the incorporation of organic manure and crop straw into paddy soils can increase both the production of rice and soil fertility (Mappaona and Yoshida, 1994; Campbell et al., 1991;

Gagnon et al., 1977; Soon, 1998; Gill and Meelu, 1982; Muhammad et al., 1992). However, many livestock and crop farms in China mismanage their animal wastes and crop straw, allowing the loss of valuable nutrients and causing many environmental problems such as water and air pollution when these resources are piled up or burned.

Phosphorus (P) is an essential nutrient element required by plants, and is an important environmental factor controlling crop growth and productivity as a result of its effects on photosynthesis and metabolism (Rodriguesz et al., 1998; Hart and Greer, 1998). Recent studies have shown that P exerts strong effects on photosynthesis. In particular, P deficiency inhibits plant growth by reducing photosynthesis per unit leaf area (Jacob and Lawlor, 1991; Kirschbaum and Tompkins, 1990; Clarkson et al., 1983). In China, soils with a P deficiency account for one-third of the total area of cultivated land (Lu, 1989).

Soil P levels can be increased by the application of chemical fertilizer, but most of the P in fertilizer that enters the soil is strongly fixed by iron and aluminum oxides, which decrease its availability (Mokwunye et al., 1986; Warren, 1992). Moreover, excessive use of chemical fertilizers can create environmental problems such as eutrophication of water and energy waste due to the energy cost of producing these fertilizers. Organic inputs generally cannot provide sufficient P for crop growth owing to their low P concentrations (Plam, 1995), but organic inputs can nonetheless increase the availability of P to some extent (Iyamuremte and Dick, 1996). Organic anions formed by the decomposition of these organic inputs may compete with P for the same adsorption sites, thereby increasing P availability in the soil (Easterwood and Sartain, 1990; Hue, 1991). Combining organic and inorganic nutrient sources may thus provide a more efficient use of scarce resources for maintaining or increasing a soil's P supply and allowing more complete utilization of soil P by plants.

The objectives of the present study were to study the dynamic characteristics of available P in soils, leaf photosynthesis, P absorption by rice plants, root morphology of these plants, and acid phosphatase activity (APA) at the root surface in response to different nutrient treatments. Our goal was to reveal the possible mechanisms responsible for the coordination of soil P supplies with rice photosynthesis.

2. Materials and Methods

2.1. Experiment site

This field experiment was conducted at the Changshu Agroecological Experiment Station (CAES) of the Chinese Academy of Sciences, a station located in the Yangtze River delta (at 31° 33′ N and 120° 42′ E) in China's Jiangsu province. The 4-yr experiment took place from 1999 to 2002. The soil in the study plots belonged to the category of gleyed paddy soils developed on lacustrine deposits. Selected soil physical and chemical properties are given in Table 1.

2.2. Experimental Design and Nutrient Treatments

The experiment was laid out in a randomized complete block design using five replications. Plot size was 5×6 m^2. The nutrient regimes were as follows: (i) control (CK), (ii)

Table 1 Selected physical and chemical characteristics of surface (0-20 cm) soil used in study

Clay	Sand	CEC[a]	pH	Organic matter	CaCO$_3$	Fe[b] OX	Al[b] OX	Total P	Available P[c]
—— % ——		cmol kg^{-1}	H$_2$O 1:1	%	—— g kg^{-1} ——			mg kg^{-1}	
46.7	15.4	16.9	7.85	3.22	113.2	13.15	5.47	0.55	8.26

[a] 1 M NH$_4$OAc method (Thomas, 1982)
[b] OX denotes acid ammonium oxalate extraction (Hodges and Zelazny, 1980)
[c] 0.5 M NaHCO$_3$ extraction (Olsen et al., 1954)

exclusively chemical fertilizers (CF), (iii) combination of chemical fertilizers with farm yard manure (FYM; CM), and (iv) combined application of chemical fertilizers and wheat straw (CS). Each plot received an equal application of 150 kg N ha^{-1}, 20 kg P ha^{-1} and 37 kg K ha^{-1}. The P and K fertilizers were basal applications of superphosphate and potassium sulfate at the time of transplanting. Nitrogen was split into three applications: 50% of the total N was provided as NH$_4$HCO$_3$ at the time of transplanting, 20% was supplied as urea at the time of tillering, and the remaining 30% was applied as urea at the time of booting. Half the dose of N in the CM and CS treatments was supplied in the form of FYM and straw, respectively, incorporated into the flooded paddy soil 2 weeks before transplanting, with the remainder provided in the form of chemical fertilizers. The total N, P, and K contents of the FYM were 19.4, 7.45, and 18.6 g kg^{-1}, respectively; wheat straw contained 3.40 g N kg^{-1}, 0.39 g P kg^{-1}, and 8.94 g K kg^{-1}.

2.3. Soil and Plant Sample Collection and Analyses

Soil samples were collected from a depth of 0 to 20 cm during specific growth stages: tillering (TS), booting (BS), heading (HS), early grain-filling (EFS), middle grain-filling (MFS), and late grain-filling (LFS). After air-drying, samples were powdered and sieved to pass through a 2-mm mesh, then were analyzed for available P using the NaHCO$_3$ method (Olsen et al., 1954). Plant samples were collected at the corresponding stages. Plant material was dried in an oven at 65 °C for 48 h before heat-processing at 105 °C for 30 min, then was ground and sieved to pass a 0.5-mm screen, and digested in hot sulfuric acid using salicylic and H$_2$O$_2$ as additives for the analysis of total P. At harvest, grain yield and various yield components were determined by oven-drying, as described above.

2.4. Leaf Photosynthetic Rate

The photosynthetic rates of the terminal leaflets of the 2nd and 3rd trifoliate leaves from the uppermost position on the plant were determined in all treatments by using a portable photosynthesis analyzer (model CI-301PS, CID Co. Ltd.,).

2.5. Acid Phosphatase Activity at the Root Surface

At specific growth stages, plants were harvested, their roots were washed with distilled water then blotted dry, and 70-mm portions of the root tip were placed directly into (typically) 5.0 mL of substrate (2.5% (w/v) p-nitrophenyl phosphate in 50 mM sodium citrate buffer, pH 5.6) and incubated at 30 ℃ for 30 min. To ensure linearity, 50 μL aliquots of the substrate reaction mix were removed at 20-min intervals for 1 h, each was added to 50 μL of 1.0 M NaOH, and the absorbance was read at 405 nm on an Anthos HTTII Plate Reader. Root morphology (area) was determined by using Zhang's (1999) method.

2.6. Statistics and Data Analysis

The data were analyzed statistically by ANOVA to determine significant differences and by Duncan's Multiple Range Test (DMRT) to separate the means. All statistical analysis was performed with the SAS software (SAS Institute, 1995).

3. Results and Discussion

3.1. Soil Phosphorus Supply

The characteristics of the soil P supply differed significantly among the different nutrient

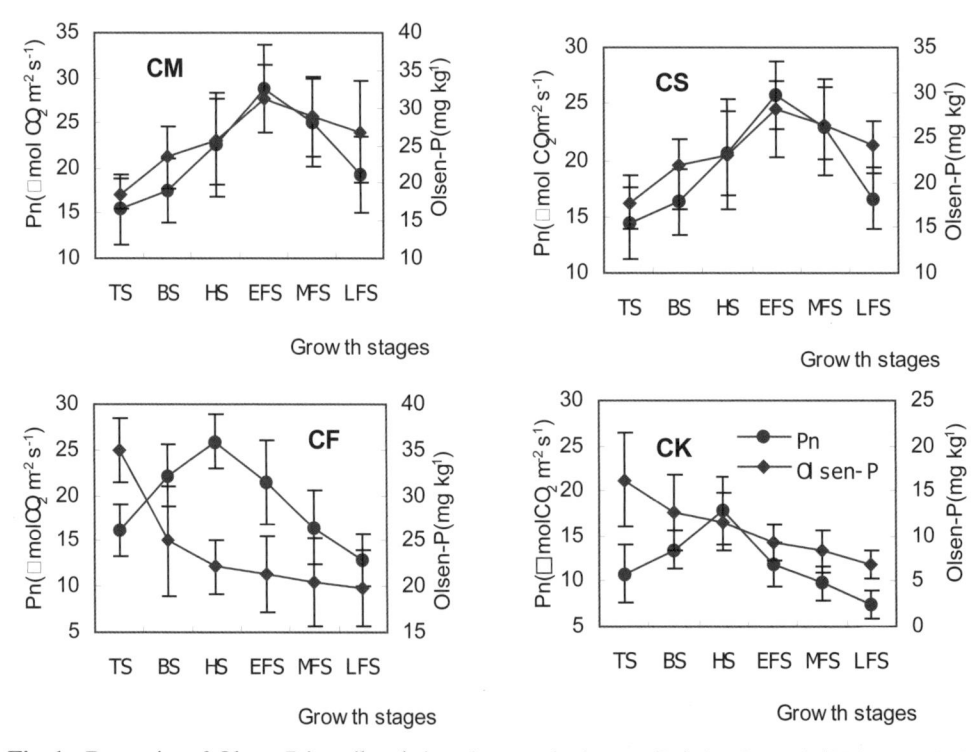

Fig.1 Dynamics of Olsen-P in soil and rice photosynthetic rate (Pn) for Control (CK), chemical fertilizers (CF), chemical fertilizers + wheat straw (CS) and chemical fertilizers + farm yard manure (CM) treatments. TS,BS,HS,EFS,MFS,LFS indicate tillering, booting, heading, early filling, middle filling and latter filling stage, respectively

treatments at different growth stages of the rice plants (Fig. 1). Levels of soil P in the chemical fertilizer (CF) treatment decreased continuously as the rice grew and developed, reaching a minimum of 18.8 mg kg^{-1} at the middle and late grain-filling stages. In the CS treatment, soil P increased after the heading stage (HS), but the increase was so small that at the grain-filling stage (FS), soil P was only 25.8 mg kg^{-1}. In the combined application of chemical fertilizers and FYM (CM), the dynamics of soil P were relatively complicated. During the early growth stages, soil P increased until around the early grain-filling stage (EFS), then subsequently decreased; this resulted in greater available P in the soil than in the other treatments, averaging 30.7 mg kg^{-1} during the grain-filling stages. This resulted from a net release of P from the organic manure incorporated into the soil, and also from a decrease in the soil's capacity to adsorb P due to competition for P-adsorption sites by the organic anions produced by decomposition of the organic manure (Easterwood and Sartain, 1990; Ohno and Crannell, 1996). These results showed that the combination of chemical fertilizers with organic materials, and specifically with FYM, could improve the soil's P supply.

3.2 Photosynthesis

As the rice grew and developed, the photosynthetic rate of its leaves (Pn) showed similar

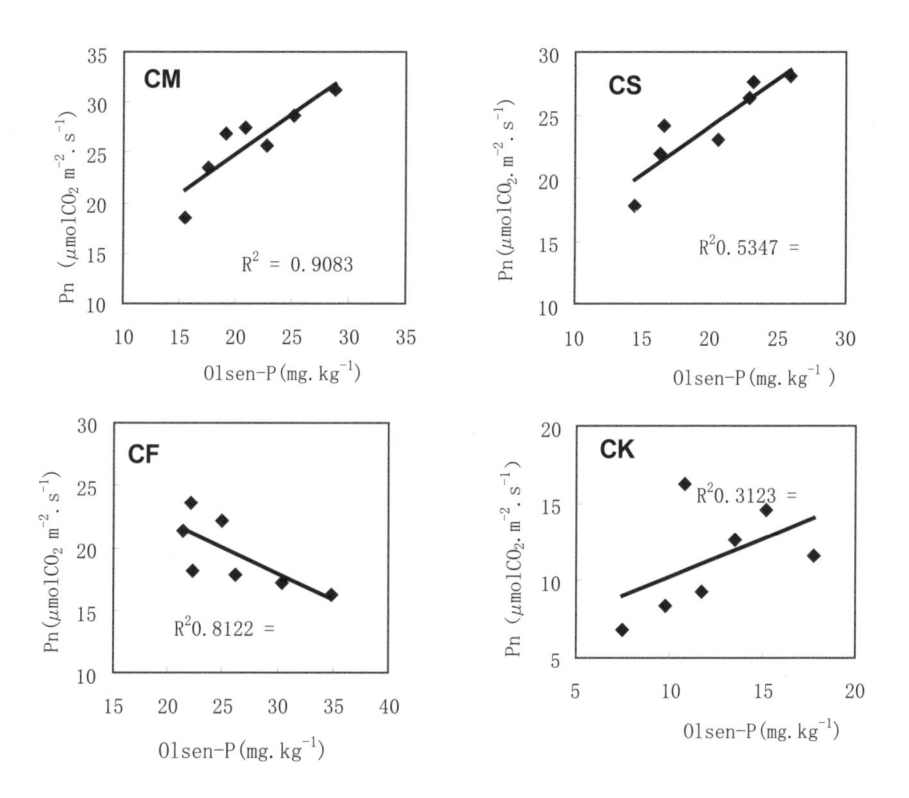

Fig. 2 Correlations of rice photosynthetic rate (Pn) to soil Olsen-P under different nutrient treatments. CK, CF, CS, CM denote control, sole chemical fertilizers, chemical fertilizers + wheat straw and chemical fertilizers + farm yard manure, respectively

trends in all treatments (Fig. 1): an initial rise, followed by a decrease. However, the patterns of photosynthetic rates and their changes differed noticeably among nutrient treatments. In the CF treatment, the photosynthetic rate increased rapidly until the booting stage, but declined so sharply after heading that Pn was only 16.5 μ mol CO_2 m^{-2} s^{-1} at the middle grain-filling stage. In the CM treatment, Pn increased slowly during the early growth stages, and for a longer period than in the CF treatment (until the EFS stage); moreover, the subsequent decrease was smaller, and the leaves maintained higher Pn even during the middle to late grain-filling stages, averaging 26.2 μ mol CO_2 m^{-2} s^{-1}. In the CS treatment, the trend in Pn was similar to that in CM, though Pn declined more rapidly than in CM around the middle grain-filling stage (MF).

By comparison, it was clear that the extent to which the available soil P coordinated with leaf photosynthesis differed among treatments. In the CM treatment, the dynamics of available soil P were more closely synchronized with Pn than in the other nutrient treatments, especially after the booting stage.

Correlation analysis between soil P and Pn for the different nutrient treatments is shown in Figure 2. In the CM treatment, Pn was significantly and positively correlated with soil P after the booting stage ($R^2 = 0.9083$, $P < 0.01$), whereas in the CF treatment, Pn was significantly and negatively correlated with soil P ($R^2 = 0.8122$, $P < 0.05$). In the CS and CK treatments, Pn was correlated positively but not significantly with soil P ($R^2 = 0.5347$). These results further revealed that the degree of coordination between available soil P and Pn was largest in the CM treatment, which combined inorganic fertilizers with organic manure.

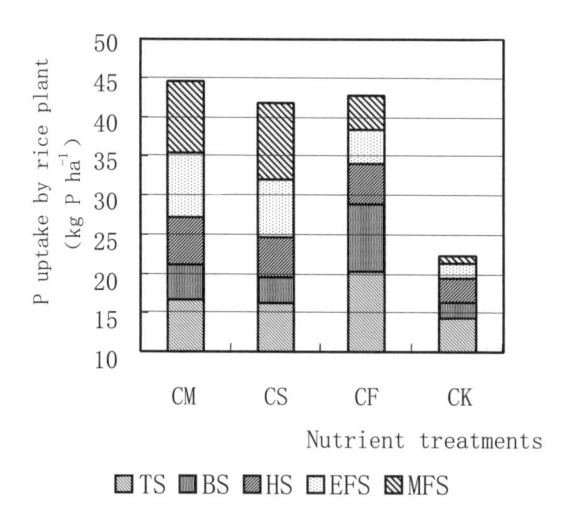

Fig. 3 P uptake by rice plant with rice growth for Control (CK), chemical fertilizers (CF), and chemical fertilizers + wheat straw (CS) and farm yard manure (CM) treatments. TS, BS, HS, EFS, MFS, LFS indicate tillering, booting, heading, early filling, middle filling and latter filling stage, respectively

3.3. Phosphorus Uptake by Rice Plants

Uptake of P by rice plants at different growth stages in the four nutrient treatments is shown in Figure 3. Although no significant ($P < 0.05$) differences were detected among the three main treatments in total P uptake by the rice plants, uptake was significantly lower in the control (CK). The distribution of the P uptake throughout the various growth stages differed significantly among the nutrient treatments ($P < 0.05$). In the CF treatment, the majority (65.8%) of the P absorbed by the plants was absorbed during the early growth stages, whereas only small amounts of P were taken up during the middle and late growth stages. In the CS treatment, the proportion of the P absorbed during the early or middle growth stages remained relatively low, but increased slightly at the middle grain-filling stage (MFS). In contrast, the combined application of chemical fertilizer and FYM (CM) showed a greater proportion of P uptake from the booting to the filling stages; 62.5% of total P absorption occurred during these stages, and especially during the middle and late grain-filling stages, the rice plants maintained a high level of P absorption.

3.4. Root Morphology and Acid Phosphatase Activity at the Root Surface

Acid phosphatase, which is present at the root surface, is believed to be responsible for the hydrolysis of organic P in the soil. Therefore, the activity of this enzyme is a significant indicator of P cycling and plant nutrition (Dick et al., 1983; Sun and Zhang, 1992). The results of the present study reveal that organic nutrient sources, particularly FYM and crop straw, promoted acid phosphatase activity (APA), which was beneficial for the rice plants because this change increased P absorption and helped maintain high Pn after booting.

The responses of APA at the root surface in the various nutrient treatments appear in Figure 4. The patterns of change in root-surface APA differed significantly ($P < 0.05$) among the treatments. In the CF treatment, the root-surface APA decreased consistently after booting, and dropped even more sharply 14 days after booting; APA averaged only 0.678 mg

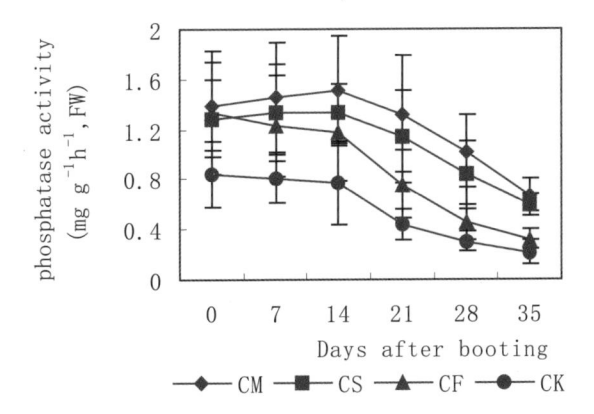

Fig. 4 Comparisons of root surface acid phosphatase activity (APA) of rice plants among nutrient treatments. CF, CS, CM, CK denote sole chemical fertilizers, chemical fertilizers + wheat straw, chemical fertilizers + farm yard manure and control, respectively

Table 2 Root morphology of rice plants under the nutrient treatments

Nutrient treatments	Root length (cm)[b]	Root dry weight (g plant^{-1})[b]	Root absorption area (m^2 plant^{-1})[b]	Root density (cm cm^3)[b]
CM	27.13 a	18.81 a	11.21 a	3.74 a
CS	25.21 b	17.32 a	9.85 b	3.01 b
CF	21.45 c	16.06 b	8.55 c	1.91 c
CK	17.86 d	13.20 c	6.15 d	0.78 d

[a] CF, CS, CM, CK indicate chemical fertilizers, chemical fertilizers+straw, chemical fertilizers+farm yard manure treatments and control, repectively.
[b] Not significantly different when followed by the same letter in columns

g^{-1} h^{-1} from 14 to 35 days after booting. In the CM and CS treatments, root-surface APA continued to increase until about 14 days after booting, after which it decreased; the ranges in the values and the rates of decrease were also much smaller than in the CF treatment. In the CM and CS treatments, root-surface APA averaged 1.132 and 0.928 mg g^{-1} h^{-1} from days 14 to 35 after booting, respectively. The CM treatment was most effective in increasing root-surface APA after the booting stage. These results demonstrate that the integration of organic materials with inorganic fertilizers could dramatically improve the root-surface APA of rice plants during the middle and late growth stages. This would contribute to maintaining high leaf Pn and high amounts of P uptake by the rice plants during the grain-filling stages.

The root morphology of the rice plants was also significantly ($P < 0.01$ or $P < 0.05$) affected by the nutrient treatments (Table 2). CM treatment produced the greatest values of root length, root dry weight, and root density at the grain-filling stage, followed by the CS treatment. The differences in root absorption area also differed significantly ($P < 0.01$). These results suggest that the combined application of inorganic fertilizers and either FYM or crop straw promoted the development of fine roots and a large root absorption area, even in the late growth stages.

3.5. Rice Yield and Yield Components

The rice yield and the various yield components were also significantly influenced by the

Table 3 Comparison of yield and yield components of rice under different nutrient treatments

Nutrient treatments[a]	No. full grain.panicle^{-1}	Grain yield (%)	Full grain 1000 (g)	grains weight (kg.hm^{-2})
CF	138.4	80.2	25.95	7543.2
CS	141.7	83.1*	27.23*	8082.9*
CM	142.9[b]*	84.5**	23.79**	8207.8**
CK	113.4	80.9	25.01	6412.8

[a] CF, CS, CM, CK indicate chemical feritilizers, chemical fertilizers+straw chemical fertilizers+farm yard manure treatments and control, respectively.
[b] *, ** denotes that treatment differences were significant at P<0.05, 0.01, respectively

nutrient regimes, as shown in Table 3. The combination of organic (FYM or straw) and inorganic nutrient sources significantly ($P < 0.05$ or $P < 0.01$) increased the productivity of the rice plants. In the CS and CM treatments, rice yields were 11.6% and 13.1% higher than in the chemical fertilizer (CF) treatment, respectively, and the ratios of full grains and of grain weights also differed significantly among the treatments ($P < 0.05$ or $P < 0.01$), suggesting that the CM treatment increased rice yields mainly by increasing these two ratios.

4. Conclusions

The integration of organic materials with inorganic sources of nutrients is often assumed to provide additional benefits compared with the application of inorganic nutrients alone. In the current experiment, the combination of organic manure and chemical fertilizers (the CM treatment) significantly increased both the soil's available P and Pn during the middle to late growth stages of the rice plants. This contribution also resulted in closer coordination of soil P with Pn than in the treatment with inorganic fertilizers alone (CF).

Acknowledgments

This study was financially supported by the National Key Project (No. 1999011802) and by the Chinese Academy of Sciences (project KZCX2-413). The authors thank the Changshu Agroecological Experiment Station of the Chinese Academy of Sciences for providing the study site and considerable help in performing the experiment.

References

Amrani, M., Westfall, D. G. and Moughli, L. (2001) Phosphorus management in continuous wheat and wheat-legume rotations. *Nutrient Cycling in Agroecosystems* 59: 19-27

Barber, S. A. (1979) Corn residue management and soil organic matter. *Agron. J.* 71: 625-627

Beck, M. A. and Sanchez, P. A. (1995) Soil phosphorus fraction dynamics during 18 years of cultivation on a typical Paleudult. *Soil Sci. Soc. Am. J.* 58: 1424-1431

Campbell, C. A., Lafond, G. P., Zentner, R. P. and Biederbeck, V. O. (1991) Influence of fertilizer and straw baling on soil organic matter in a thin Black Chernozem in western Canada. *Soil Biol. Biochem.* 23: 443-446

Clarkson, D. T., Kerridge, P. C., Sherriff, D. W. and Fisher, M. J. (1983) Effects of phosphorus deficiency on photosynthesis and stomatal function in the tropical legume, Siratro. Ann. Rep. ARC Letcombe Lab.: 65-66

Constant, K. M. and Sheldrick, W. F. (1991) An outlook for fertilizer demand, supply, and trade, 1988/89-1993/94. World Bank Technical Paper No. 137, The World Bank, Washington, D. C.

Daniel R., Keltjens, W.G. and Goudriaan, J. (1998) Plant leaf area expansion and assimilate production in wheat growing under low phosphorus conditions. *Plant Soil* 200: 227-240

Dick, W.A. and Tabatabai, M. A. (1983) Effects of soil on acid phosphatase and inorganic pyrophosphatase of corn root. *Soil Sci.* 136: 19-25

Easterwood, G.W. and Sartain, J.B. (1990) Clover residue effectiveness in reducing orthophosphate sorption on ferric hydroxide coated soil. *Soil Sci. Soc. Am. J.* 54: 1345-1350

Gagnon, B., Simard, R.R., Robitaille, R., Goulet, M. and Rioux, R. (1977) Effect of composts and inorganic fertilizers on spring wheat growth and N uptake. *Can. J. Soil Sci.* 77: 487–495

Generose, N., Cheryl, A. P., Roland, J. B. and Paul, C. S. (1998) Soil phosphorus fractions and adsorption as affected by organic and inorganic sources. *Plant Soil* 198: 159–168

Gill, H. S. and Meelu, O.P. (1982) Studies on the substitution of inorganic fertilizers with organic manure and their effect on soil fertility in rice wheat rotation. *Fertil. Res.* 3: 303–313

Hart, A. L. and Greer, D. H. (1988) Photosynthesis and carbon export in white clover plants growth at various levels of phosphorus supply. *Physiol. Plant.* 73: 46–51

Hodges, S. C. and Zelazny, L. W. (1980) Determination of noncrystalline soil components by weight difference after selective dissolution. *Clays Clay Miner.* 28: 35–42

Iyamuremye, F. and Dick, R. P. (1996) Organic amendment and phosphorus sorption by soil. *Adv. Agron.* 56: 139–185

Jacob, J. and Lawlor, D. W. (1991) Stomatal and mesophyll limitations of photosynthesis in phosphate deficient sunflower, maize and wheat plants. *J. Exp. Bot.* 42: 1003–1011

Kirschbaum, M. U. and Tompkins, D. (1990) Photosynthetic responses to phosphorus nutrition in Eucalyptus grandis seedlings. *Austral. J. Plant Physiol.* 17: 527–535

Lur, K. (1989) The basic conditions of soil nitrogen, phosphorus and potassium in China. *Acta Pedologica Sinica* 26: 280–286

Luo, A. and Sun, X. (1994) Effect of organic manure on the biological activities associated with the release of phosphorus in blue purple paddy soil. *Commun. Soil Sci. Plant Anal.* 25: 2513–2522

Mappaona, I., Shigekata Y. and Makoto, K. (1994) Yield response of cabbage to several tropical green manure legumes incorporated into soil. *Soil Sci. Plant Nutr.* 40: 415–42

Muhammad, I.C. and Kounosuke, F. (1998) Comparison of phosphorus deficiency effects on the growth parameters of mashbean, mungbean, and soybean. *Soil Sci. Plant Nutr.* 44: 19–30

Muhammad, S.Z., Muhamad, M., Muhammad, A. and Maqsood, A.G. (1992) Integrated use of organic manure and inorganic fertilizers for the cultivation of lowland rice in Pakistan. *Soil Sci. Plant Nutr.* 38: 331–338

Ohno, T. and Crannell, B.S. (1996) Green and manure-derived dissolved organic matter effects on phosphorus sorption. *J. Environ. Qual.* 25: 1134–1143

Olsen, S.R., Cole, C., Watanabe, F.C. and Dean, L.A. (1954) Estimation of available phosphorus in soil by extaction with sodium bicarbonate. USDA, Washington, D.C. Circular: 939 pp

Rao, D.N. and Mikkelson, D.S. (1976) Effect of rice straw incorporation on rice plant growth and nutrition. *Agron. J.* 68: 752–755

SAS Institute (1995) SAS/Stat User Guide. Vol. 2,Version 6.1. SAS Inst., Cary N. C.

Singh, B.B. and Jones, J.P. (1976) Phosphorus sorption and desorption characteristics of soil as affected by organic residues. *Soil Sci. Soc. Am. J.* 40: 389–394

Soon, Y. K. (1998) Crop residue and fertilizer management effects on some biological and chemical properties of a Dark Grey Solod. *Can. J. Soil Sci.* 78: 707–713

Sun, X. and Zhang, Y.S. (1992) Effects of organic phosphorus in organic manures and paddy soil on rice growth. *Acta Pedologica Sinica* 29: 365–369

Terry, N. and Rao, I. M. (1991) Nutrients and photosynthesis: iron and phosphorus as case studies. In Porter, J.R., and Lawlor, D.W. (Eds.), Plant Growth Interactions with Nutrition and Environment. Soc. Exp. Biol., Seminar Series 43. Cambridge University Press, Cambridge: 55–79

Thomas, G. (1982) Exchangeable cations. In Page, A.L (Eds.), Methods of soil analysis. Part 2., Agro. Monogr. 9, Madison:159–165

Warren, G. (1992) Fertilizer phosphorus sorption and residual value in tropical African soil. NRI Bulletin 37. Natural Resources Institute, Chatham, England.

Zhang, Z. (1999) Experiment guide for plant physiology. High Education Press, Beijing, China.

Development of a comprehensive system for the analysis and evaluation of water quality in medium-sized watersheds

Sunao Itahashi and Makoto Takeuchi

National Institute for Agro-Environmental Sciences, 3-1-3 Kannondai, Tsukuba, Ibaraki 305-8604, Japan

---

Abstract

A Windows-based system including a simulation model was developed for the analysis of water quality, especially nitrogen levels in river water. The model requires geographical data such as land use, contours, and surface soils, as well as data from the Agricultural Census, sewer distribution data, meteorological data, and pollutant load factors for sources of nitrogen load emission. The model generates annual nitrogen outflow potentials within the catchment that drains to user-specified location on a digitized map. The model also takes account of removal of nitrogen through denitrification in flooded zones such as rice paddies. Such removal was quantitatively evaluated on the basis of land-use chain status. The system's multivariate analysis tools facilitates to analyze and display the relationships between nitrogen concentrations in a river water and data of its fluctuating factors such as daily precipitation by incorporating those factors as variants in a regression formula, effectively clarifying variations in monitored data and estimating annual nitrogen load in river water. Applying the model to several watersheds in Japan adequately estimated the soil surface nitrogen balance and the nitrogen discharge in the water, revealing a time lag between nitrogen loading and nitrogen discharge appeared in aquatic environments. When water quality in wells was analyzed using land-use patterns of each well's catchment, nitrogen input was shown to derive from water from the catchment as well as from soils near the well.

Key words : Denitrification, land-use chain, nitrogen, Windows-based simulation model, catchment

---

1. Introduction

Nitrogen is an element that sensitively reflects anthropogenic contamination of rivers and other aquatic environments. Because agricultural land receives large inputs of nitrogen in inorganic fertilizer and manure to stimulate crop production, it has been recognized as a major non-point source of pollution (Kunimatsu, 1989). However, it is difficult to tell how much of the nitrogen applied to agricultural lands actually reaches aquatic environments, and when, because water and nitrogen movements in the soil are complicated processes.

Currently, local governments in most prefectures in Japan have been using a simple "pollutant load factor" (PLF) method to predict and evaluate water quality while developing pollutant-control plans (e.g., Harasawa, 1990). However, the method is not capable of answering the questions of how much nitrogen enters the water and when (Nakanishi, 1985); moreover, concerning environmental pollution, it neglects the important "denitrification" function served by flooded areas such as marsh lands and especially rice paddies, which are a typical, widespread feature of monsoon Asia.

The Organization for Economic Cooperation and Development (OECD) has proposed the use of risk indicators for water quality to evaluate the effects of agricultural practices on water pollution (OECD, 2001). However, the proposed method considers only the soil surface nitrogen balance and water balance in the calculation of potential nitrogen concentrations in agricultural soils, and does not consider water and nutrient movements into nearby aquatic environments through the soil. This also means that like the simple PLF method, the OECD method neglects the effects of denitrification.

To overcome the weak points of both methods, we have been developing a method capable of more precise predictions of nitrogen dynamics in any watersheds. We have developed a prototype system that runs under Microsoft Windows and includes a simulation model for the evaluation of water quality. The model combines various kinds of geographical and numerical information as inputs in order to calculate both the nitrogen outflow potentials and the states of a "land-use chain (a cascade-like concept on land-use transitions defined by water movement)". The model can also simulate actual nitrogen discharges into a river, and provides tools for multivariate analysis that facilitate to generate multivariate regression formulas of monitored water flows and nitrogen concentrations, and to analyze the relationships between the monitored data and various fluctuating factors, especially climate, revealing the effects of those factors on the variation of the data.

The present paper discusses some examples of our model's outputs.

2. Outline of the simulation model

2.1 Estimation of nitrogen loading in a catchment

Our model adopts the PLF method (e.g., Nakanishi, 1985, and Ukita and Nakanishi, 1989) to calculate the nitrogen loads and modified it to estimate outflow potential from a body of land, which is divided into individual units of land-uses such as forested land, agricultural land (paddy field, upland field, grass land, and orchard), residential areas, aquatic area, and bare land (none of the above). The following nitrogen sources are considered in the model: precipitation, fertilizer and manure applications on agricultural land, human excreta, and "gray water (e.g., wash water from human residences)". Details of the processes for estimating the nitrogen loading and outflow potential are described in Takeuchi (1992).

Table 1 Contents in the simulation model for analysis and evaluation of water quality (updated from Takeuchi, 1992)

Contents	Processed Data
Geographical data digitalization	Land use, Contour, Surface soil, Sewer distribution, Administrative district (new and old), Basin outline, etc.
Remote objects capture	Same as above, AMeDAS data and *SHP* type files
Table data digitalization and conversion to basin based data	Agricultural census, Sewer distribution rates, AMeDAS, PLFs of fertilizer application and Nutrient uptake efficiency, etc., Decomposition property of organic materials
Land use chain status of water movement	Mesh data of altitude developed by Land use, River, and Contour maps
Dynamics of nitrogen in organic matter	Carbon and nitrogen composition of organic materials, Application rates, AMeDAS, Statistical Data Book of Soil Survey
Water quality data analysis	Monitoring data of water quality and quantity, AMeDAS
Water quality prediction in a river	(All data)

2.2 Computer programs

All computer programs in the system were written and compiled using Borland C++ for Windows (Borland International Inc.; http://www.borland.co.jp/). Table 1 describes the contents of the system. Additional details on each item are presented by Takeuchi (1992) and later in the present paper.

2.3 Digitizing of geographical data

Printed land-use and contour maps at a scale of 1:25000 (published by Japan's Geographical Survey Institute, http://www.gsi.go.jp/) and maps of the surface soil type at a scale of 1:50000 (published by local governments administered by National Land Agency, or currently National Land Survey Division, Land and Water Bureau, Ministry of Land, Infrastructure and Transport, http://tochi.mlit.go.jp/tockok/index.htm) were digitized using a scanner to produce bitmap files. The data were then processed using a self-produced PC program included in the system and standardized so that individual pixels represented square areas of about 13 m × 13 m in size. The scale of these pixels reflects the realities of Japanese land-use patterns.

2.4 Utilization of external digital data

Our system can incorporate and process existing digital data in a range of formats, such as *SHP* files (commonly used by GIS software such as ArcInfo; ESRI, http://www.esri.com/), and water quality monitoring data files compiled by Japan's National Institute of Environmental Studies and published by the Environmental Information Center (http://www.eic.or.jp/eic/).

2.5 Digitizing of tabular data and conversion into catchment-based data

The model requires various printed tabular data such as the Agricultural Census data compiled by the Statistics and Information Department of Japan's Ministry of Agriculture, Forestry and Fisheries and published by Norin Tokei Kyokai (http://www.aafs.or.jp/); AMeDAS meteorological data compiled by the Japan Meteorological Agency and published by the Japan Meteorological Business Support Center (http://www.jmbsc.or.jp/main_html/index1.htm); rates of fertilizer application and recovery by crops (e.g., Ogawa, 2000); the distribution of municipal sewers, and so on. Those data are manually input into any spreadsheet software compatible with the *CSV* file format, including Microsoft Excel (Microsoft, http://www.microsoft.com/office/). The system can convert municipal-based data into catchment-based data by calculating nitrogen balances for individual pixels when digitized geographical maps are successfully processed and the extent of the catchment is properly determined (see 2.3 and 2.6 also).

2.6 Land-use chain and water flow

The direction of movement of water within a pixel is determined by evaluating topographical information from the digitized contour maps; that is, the altitudes of all pixels are calculated from the contour maps using self-produced program. It is assumed that water always flows in the most strongly downward direction until it reaches nearby water body. The system also calculates the extent of catchment that drains to the user-specified location on the digitized map by accumulating the outputs of the water-flow function.

2.7 Dynamics of nitrogen in organic matter

The function of nitrogen in organic matter has been modified from the original model (Takeuchi, 1992). The amounts of livestock feces and urine production within an administrative district are estimated from the number of beef and dairy cattle, pigs, and layers and broilers shown in the Agricultural Census; these numbers are used as "frame values (required fundamental variables for calculation of nitrogen loadings)", and then multiplied by the corresponding pollutant-load factors (Ukita and Nakanishi, 1989). Most of the total production of fecal wastes is accounted for on the assumption that these wastes are processed and applied to soils as farmyard manures (e.g., Statistics and Information Department of MAFF, 2001). However, significant amounts of nitrogen in the wastes are estimated to be lost through volatilization as ammonia (NH_3) during temporary storage and manure-production processes as well as through denitrification just after application (Saito, 1989; Matsumura, 1988). Nitrogen-loss ratios for cattle and pig urine were assumed to be 0.5 (Matsumura, 1988), and those for feces were assumed to be 0.23 for cattle, and 0.149 for pigs, and 0.447 for chicken droppings on the basis of comparisons of ash concentrations in feces and droppings before and after drying treatment (Ogata, 1983); uniform ash contents were assumed through the treatment.

The distribution rates of manure to agricultural fields of every crop type were allocated proportionally on the basis of the cultivation area ratios for various crop types from the

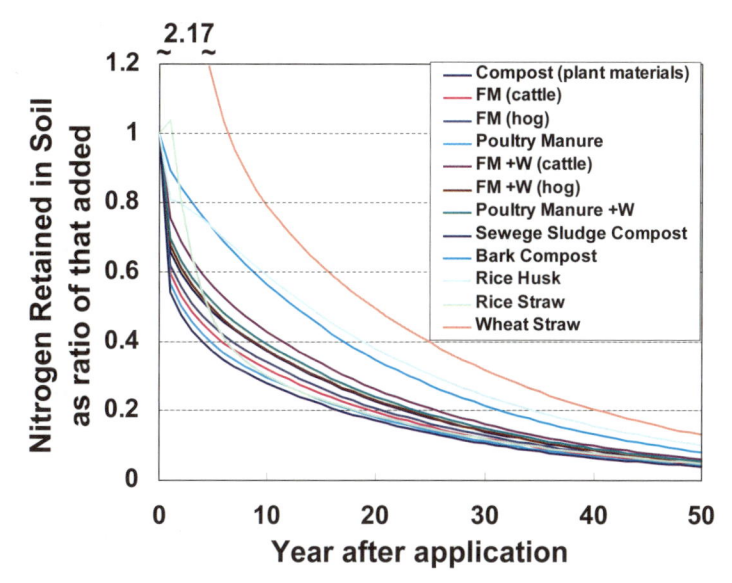

Fig. 1 Estimated nitrogen retention ratios of various organic materials in Yahagi River basin
annual accumulated temperature: 5353.5 degree Celsius
(average of 1992–1996 in Toyohashi, Aichi pref.)
FM: farmyard manure
+W: with chips of wood
Some ratios go over one, namely original concentrations, when effects of immobilization are predominant.

Agricultural Census. The same distribution rates of sewarage-sludge compost were assumed.

The dynamics of soil nitrogen derived from organic materials such as manure and compost are treated differently from those of inorganic fertilizers. Because organic material contains much more organic nitrogen than inorganic nitrogen and because organic nitrogen gradually decomposes into an inorganic form, the time required for the mineralization process must be considered when determining the actual loadings from the organic materials in a given year. Annual rates of release of inorganic nitrogen from organic material are estimated based on the carbon and nitrogen composition of the materials (Agricultural Production Bureau, 1982); three hypothetical components of organic carbon with different half-lives are considered (Inoko, 1985), and the cumulative annual temperature (in degrees) of the location under consideration is included as a model parameter (Fig. 1).

2.8 Water quality analysis

The system contains tools for combined multivariate analysis by means of multiple regression, orthogonal polynomials, and analysis of variance. Applying these tools permits various statistical analyses of the data sets and visualization of the results.

2.9 Water quality prediction for rivers

The nitrogen load at any location in a river is estimated by subtracting the estimated amount of nitrogen removed through denitrification in the flooded zone within the catchment from the total nitrogen loading in the corresponding area. To estimate the amount of nitrogen removal in the flooded zones, the amount of water flowing into the zones and its nitrogen concentration are required because nitrogen removal rates through denitrification have been reported to be proportional to the nitrogen concentrations in the influents (Tabuchi and Takamura, 1985).

Here to describe the most complicated case of rice paddies out of all flooded area. In general, the main sources of influents are precipitation, inflowing water from higher areas, and irrigation water from rivers or ground water. To estimate the total amount of water that flows into a group of rice paddies, the maximum potential water percolation during an irrigation period is estimated for each pixel constituting rice paddies by considering the "water requirement in depth" for the soil type at the location (e.g., Shiga et al., 1985). The difference between the maximum potential water requirement and the sum of inflowing waters such as precipitation, water from higher altitudes, and the minimum irrigation water (200 mm) is used as a compensation factor to be added to the amounts of irrigation water when the difference is positive.

The nitrogen concentration in irrigation water is estimated by calculating the amount of nitrogen loadings within the whole sub-catchment area that drains to the location where the water was taken, then dividing this by the amount of precipitation for the area. The inflowing nitrogen load in the inflowing water from higher areas is estimated similarly for a group of rice paddies, except that the extent of the sub-catchment is decided for the whole outline of the paddies.

3. Sample outputs from simulations using our model

3.1 Nitrogen balance in two catchments

Figure 2 shows the calculated nitrogen balance for part of the Koise River catchment (122 km^2 drainage area). Eight kinds of land use (forest land, bare land, grass land, orchard, upland field, residential area, paddy field, and aquatic area) received nitrogen from five different sources: fallout, human excreta, gray water, fertilizer, and animal excreta. The simulation results showed that agricultural fields (grass land, upland farm, orchards and rice paddies) received large amounts of nitrogen, but that only a small proportion of this input was recovered by the crops. Although the rest of the nitrogen was considered as potential loading into the river, nearly half of it was considered to have been stored in the soil, and denitrification also reduced the amount of nitrogen that actually flowed into the river.

Figure 3 shows another example of the results of a simulation study. Once all necessary numerical and geographical data have been compiled and processed, a mouse click on any location within the digitized map triggers a calculation to determine the extent of the sub-catchment that drains to the pointed location as the drain outlet, and, at the same time, to determine the nitrogen balance of each pixel in the sub-catchment. The results of these estimates of nitrogen balance are accumulated and converted into a net balance for each type of land-use. In this figure, part of the Seimei River catchment (26 km^2 drainage area) around

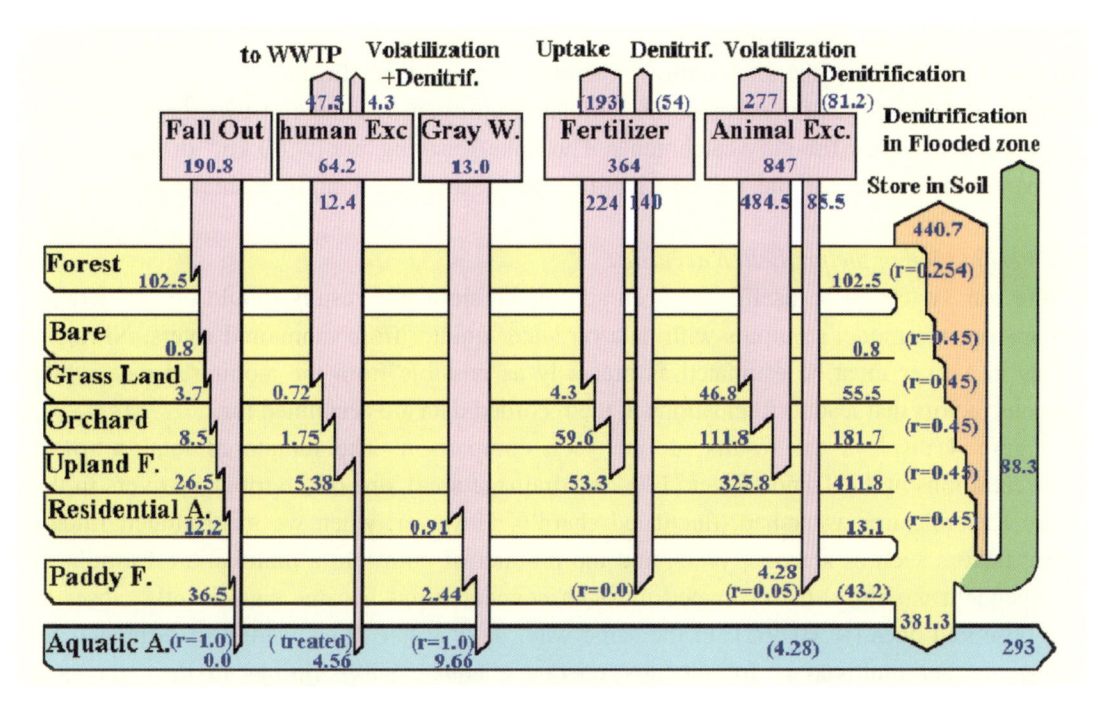

Fig. 2 Calculated nitrogen balance in Koise river basin, Ibaraki prefecture
units of nitrogen : N ton/year

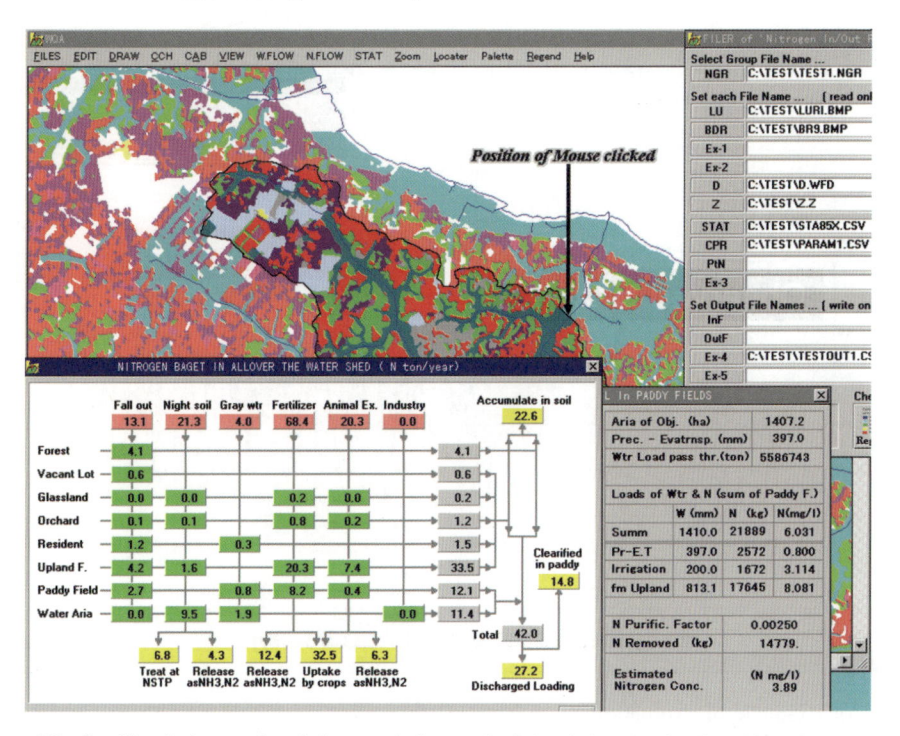

Fig. 3 Simulation study of nitrogen balances in Seimei river basin, Ibaraki prefecture
unit of nitrogen : N ton/year

Lake Kasumigaura was shown to remove significant amounts of nitrogen (14.8 t/year, equivalent to 35.2% of the total outflow potential) through denitrification.

The model has thus far successfully simulated nitrogen dynamics in two medium-sized watersheds and their tributary catchments: The Lake Kasumigaura (2200 km² drainage area) in Ibaraki prefecture, and the Yahagi River (2200 km² drainage area) in Aichi prefecture.

3.2 Verification of the prediction accuracy

For our system to be useful, the accuracy of the simulated results should be verified by comparing the model's outputs with data on water quality from monitored rivers. Nitrogen loads in a river must be estimated as precisely as possible from the monitored data while various factors that lead to fluctuations in the recorded data are accounted for.

Figure 4 displays the results of one such comparison. The monitored total nitrogen concentrations in the Tomoe River (129 km² drainage area), one of the tributary rivers in the Lake Kasumigaura watershed, fluctuated sharply. However, when we incorporated fluctuating factors such as seasons, years, and the product of both into a multivariate formula as independent variables, the estimated nitrogen concentrations became significantly closer to the observed ones (R^2=0.667). In the same way, daily water flows were also estimated by using another multivariate formula to generate adequate curve fittings to the data as a function of various factors, including precipitation, used as independent variables. Daily nitrogen loads were estimated by multiplication of the estimated nitrogen concentration and

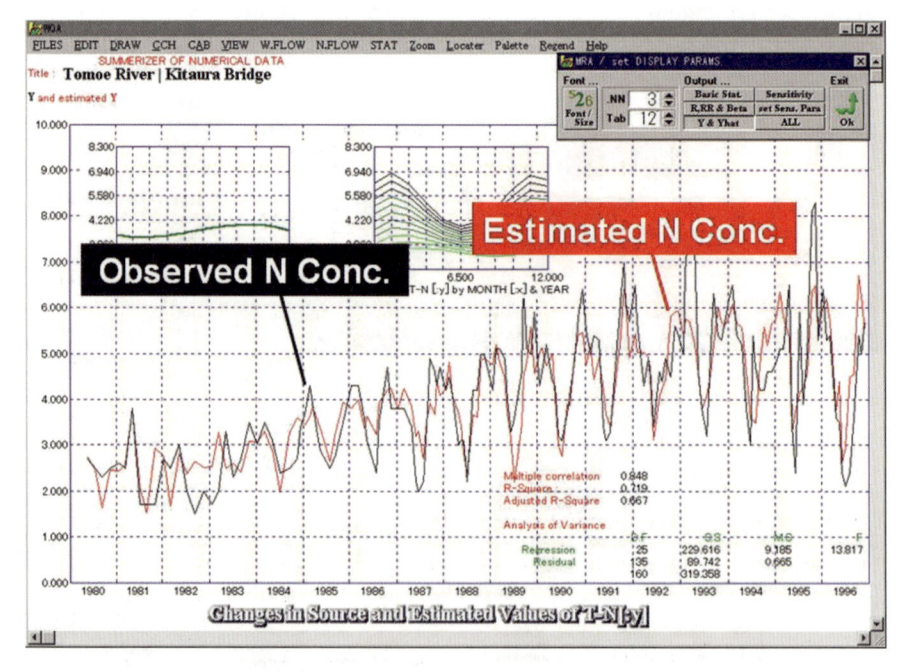

Fig. 4 An example of estimation of monitored nitrogen concentration in Tomoe river, Ibaraki prefecture, by use of a multivariate expression with season, year, and the product of both as independent variables

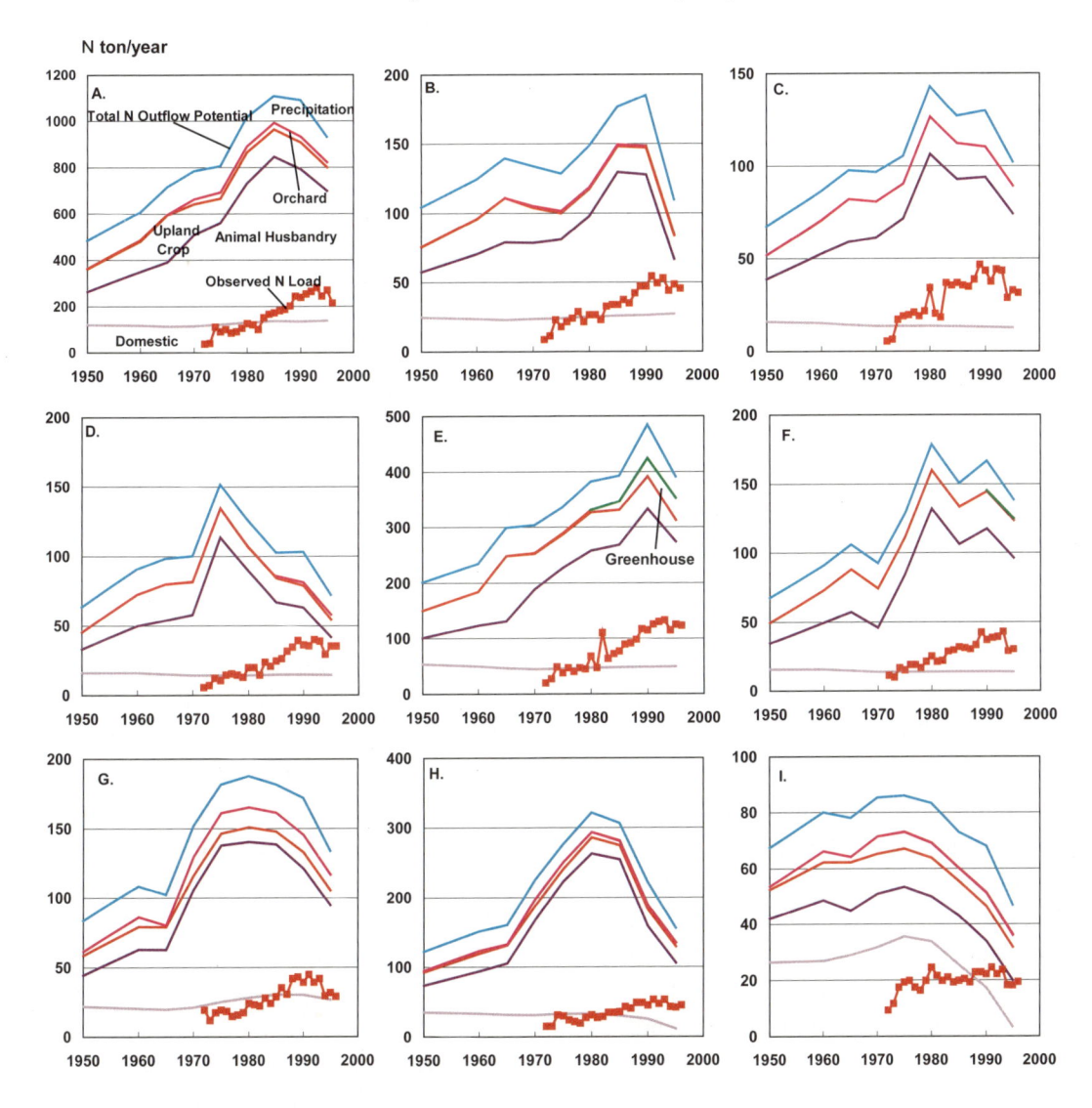

Fig. 5 Trends in nitrogen outflow potential in a river basin and estimated nitrogen load at observation site
  A. Kitaura bridge over Tomoe R. : drainage area 128.7 km^2
  B. Sekikawa bridge over Kajinashi R. : drainage area 30.6 km^2
  C. Kurakawa bridge over Kura R. : drainage area 16.7 km^2
  D. Uchiyado bridge over Takeda R. : drainage area 19.6 km^2
  E. Asahi bridge over Hokota R. : drainage area 54.1 km^2
  F. Yamada bridge over Yamada R. : drainage area 19.3 km^2
  G. Gongen bridge over Hishiki R. : drainage area 23.7 km^2
  H. Ichinose bridge over Ichinose R. : drainage area 29.5 km^2
  I. Sakai bridge over Sakai R. : drainage area 14.9 km^2

the estimated water flow on each date. Annual loads of nitrogen at the monitoring site were then estimated as the sum of the daily loads. On the other hand, nitrogen outflow potentials were estimated by the model using 5 year intervals because the Agricultural Census data were recorded every 5 years.

Estimated nitrogen outflow potentials and nitrogen loads in rivers were compared in several tributary river basins around Lake Kasumigaura (Fig. 5). Nitrogen outflow potentials gradually increased from 1950 to the 1980s, but tended to decrease sharply thereafter, owing mostly to changes in animal husbandry in these areas rather than to fluctuations in other nitrogen loads such as those from precipitation, upland crops, orchards, greenhouse culture, and domestic nitrogen loads including human excreta and gray water. Observed nitrogen loads, on the other hand, seemed to continue to increase even after the 1980s.

The differences in these trends may reflect the consequences of the integrated effects of nitrogen movements over time through the soils of the catchments (i.e., time lags) and variations in the strengths of denitrification and of water and nitrogen retention in the soils across the whole catchment, among other factors.

3.3 Quantitative analysis of nitrogen runoff ratios

Figure 6 shows a quantitative explanation of fluctuations in the "apparent nitrogen runoff ratio", which represents the ratio of the observed nitrogen loads in a river to the estimated nitrogen outflow potential in the river's catchment, based on the status of the land-use chain in the catchment. This runoff ratio reflects transitory nitrogen storage in the soil and nitrogen loss through denitrification, which may be significant in flooded zones such as rice paddies and may play an important role in reducing the actual amount of nitrogen discharged into aquatic environment. The introduction of two factors describing the land-use chain ($f1$ and $f2$; Fig. 6-b) generated sufficiently precise estimates of the nitrogen runoff ratio; the correlation coefficient between the product of $f1$ and $f2$ and the runoff ratios in five catchments equaled 0.961 (Fig. 6-c).

As shown in Fig.6-b, factor $f1$ represents the ratio of $AU1$ to the sum of $AU0$ and $AU1$, where $AU0$ represents the estimated amount of water flowing directly into the stream (out of the total amounts of water derived from the agricultural lands other than paddy fields) and $AU1$ represents the amount of water passing through the flooded zones such as paddy fields out of the same total amount of water as in $AU0$. Factor $f2$ represents the ratio of the total area of flooded zones ($A0$) to their drainage area (AT); that is, the value relates to the hydraulic retention time of water passing through the flooded zones (Kuribayashi et al., 1985). Both factors was able to be calculated by the system when all required data were successfully compiled (Fig. 6-a). This result suggested that the system was capable of quantitatively estimating apparent runoff ratios using these factors describing the land-use chain.

The results shown thus far suggest that the system is capable of estimating nitrogen loadings and evaluating the apparent runoff ratio of nitrogen at the scale of entire catchments. However, the model that predicts water quality should include functions to predict the length

a.	Basin area [Km²]	N-potential (A)	Observed N-Outflow (B)	run-off ratio (B)/(A)	f1·f2	(f1)	(f2)	fN-Removal
Kajinashi	30.6	185	53.1	0.287	0.047	0.625	0.075	0.077
Hokota	54.1	485	133.6	0.275	0.034	0.53	0.065	0.116
Yakoshi	16.7	104	22.5	0.217	0.125	0.483	0.258	0.302
Yamada	19.3	179	43.1	0.241	0.091	0.737	0.124	0.225
Takeda	19.6	152	40.2	0.265	0.066	0.549	0.121	0.148

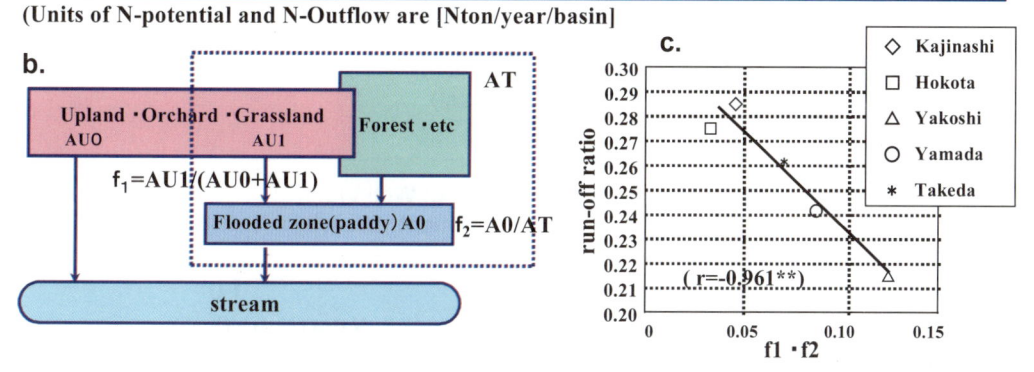

(Units of N-potential and N-Outflow are [Nton/year/basin])

Fig. 6 Quantitative analysis of apparent nitrogen runoff ratio with land use chain factors, f1 and f2, in five river basins

a. Some values concerning nitrogen removal in five river basins analyzed; '(A) N-potential' shows the peak value of estimated nitrogen outflow potential in a river basin, as shown in Fig.5, '(B) Observed N-Outflow' shows the peak value of estimated nitrogen load based on monitored data in corresponding river also shown in Fig.5, and 'fN-Removal' is calculated by one minus the ratio of the estimated runoff ratio to the potential runoff ratio which is given by assuming f1 times f2 to be zero, suggesting the ratio of nitrogen removal due to denitrification to total nitrogen load attenuation.

b. A schematic diagram of land use chain factors, f1 and f2.

c. Correlations between the products of f1 and f2 and estimated run-off ratios in five river basins.

of the time lags and the quantitative nitrogen outflows if it is to achieve adequate prediction accuracy.

Such factors as the groundwater level, water- and nitrogen-retention capacities of the soil, and the distance between the locations of nitrogen applications and outflows in a catchment were assumed to affect the length of the time lags (Nakanishi, 1985). The strengths of these factors may vary between catchments. Therefore, future research must analyze more catchments with different characteristics so as to elucidate and evaluate the factors that affect water and nitrogen movements at the scale of entire catchments.

3.4 Analysis of water quality in wells

Our last example of using the system involves an analysis of the origin of well water (Fig. 7). Nitrogen concentration data from 44 wells in the Yahagi River watershed, Aichi prefecture, were analyzed with respect to various factors, including the land-use patterns in the catchment that drains to each well. When nitrogen concentrations were analyzed using the multivariate analysis tools provided by our system, the partial regression coefficients corre-

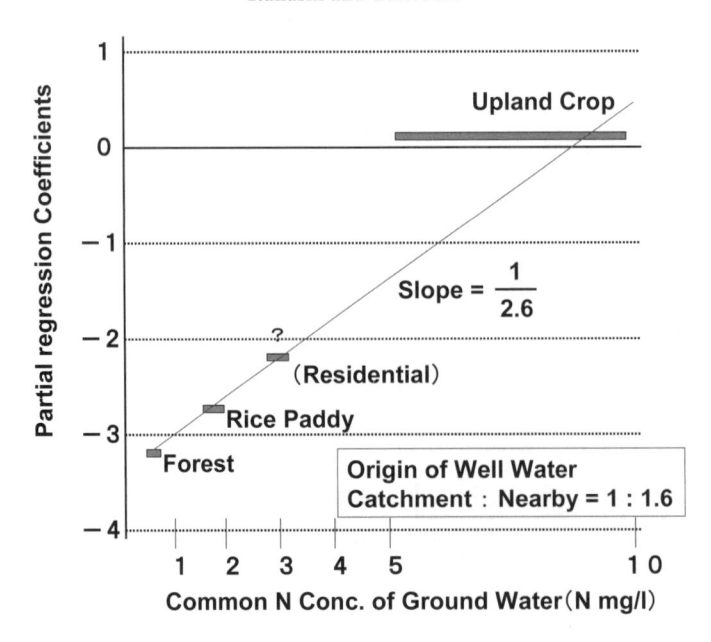

Fig. 7 Nitrogen concentrations of ground water in wells reflect land use pattern in their catchments

sponding to the proportions of the total catchment area in each land-use unit showed a linear correlation with the commonly observed nitrogen concentrations in ground water in areas with corresponding land uses. The slope of the correlation was about 1 in 2.6 (0.38), which meant that the quality of the water in a well consisted of 1.6 portions of water from near the well and 1.0 portion of water from the catchment that drained to each well. This suggests that the prediction of water quality at a specific location requires information on water movement within a catchment and on land-use patterns. This analysis may be helpful for mapping nitrogen concentrations in ground water at various locations.

4. Conclusions

The comprehensive system for simulating water quality that is described in this paper uses accumulations of all kinds of statistical and research data, combined with geographical data. Sample uses of the system showed that it is able to accurately estimate the nitrogen outflow potential in a catchment, the effects of denitrification in flooded areas, the time lags between nitrogen application and its discharge into rivers, and the water quality in wells and rivers. Additional research will be required to develop enough data on the variation between catchments to adapt the system for more widespread use.

Acknowledgments

The authors express their special thanks to members of the Water Quality Conservation Unit (Mses. Arai, Uchida, Kuroda, Okada, and Hoshino [Takada], and Dr. Komada) for their assistance for compiling the necessary data for our analyses.

References

Agricultural Production Bureau (1982) Qualities of organic materials. Ministry of Agriculture, Forestry, and Fisheries (MAFF), Tokyo, 113pp.

Harasawa, H. (1990) 1. Supporting system for watersheds managements of lakes. In: Proceedings of 3rd Symposium on Aquatic Carrying Capacity and its Application. (ed. by National Institute of Environmental Studies (NIES)), NIES, Tsukuba, Japan. pp. 1-13.

Inoko, A. (1985) Mathematical expression of decomposition and accumulation of an organic matter. In: Prediction of organic matter fluctuation and development of an organic matter application standard in agricultural lands. (ed. by Agriculture, Forestry, Fisheries Research Council (AFFRC) of MAFF), MAFF, Tokyo, Japan. pp. 34-38.

Kunimatsu, T. (1989) 1-5. Pollutant loads from agricultural lands. In: Modeling analysis of pollution in rivers. (ed. by Japanese Society of Civil Engineers), Gihodo shuppan, Tokyo, Japan. pp. 50-68.

Kuribayashi, Y., Kawashima, S., Kusuda, T., Sakiyama, M., Sato, K., Sumitomo, W., Takakuwa, T., Matsuo, T. and Baba, Y. (1985): 5. Water works and Water Quality Conservation, in Hydraulic Formulas. In: Hydraulic Formulas. (ed. by Japanese Society of Civil Engineers), Japanese Society of Civil Engineers, Tokyo. pp. 374-478.

Matsumura, S. (1988) Nitrogen volatilization from livestock urine by application to upland fields. *Jpn. J. Soil Sci. Plant Nutrition* 59: 568-572

Nakanishi, H. (1985) 2-1. Calculations of pollutant loads. In: Proceedings of Symposium on Self-Purification in Aquatic Environment -Effective Use of Natural Ecosystems for Water Quality Management. (ed. by NIES,) NIES, Tsukuba, Japan. pp. 15-27.

OECD (2001) Environmental Indicators for Agriculture. Volume 3. Methods and Results. Organization for Economic Cooperation and Development, Washington, D. C. 409 pp.

Ogata, T. (1983) Chemical composition of feces and urine of livestock, *Miscellaneous publication of the National Grass Land Research Institute*, 58-2, pp. 59-60.

Ogawa, Y. (2000) Nitrate pollution of ground water and shift of the agricultural technology. Nobunkyo, Tokyo, Japan, 200 pp.

Saito, G. (1989) Estimation of ammonia volatilization from slurry applied to grass lands and its reduction by addition of acids. *Bull. Nat. Grassland Res. Center* 41: pp. 38-48.

Shiga, H., Ohyama, N., Maeda, K. and Suzuki, M. (1985) An evaluation of different organic materials based on their decomposition pattern in paddy soils. *Bull. Nat. Agric. Res. Center*, 5: pp. 1-19.

Statistics and Information Department. (2001) An investigation on environmentally conscious agriculture. Results of livestock section. Appendix of A Report of Investigation on Status of Establishments and Managements of Composting Facilities for Livestock Wastes. MAFF, Tokyo, Japan. pp. 1-127.

Tabuchi, T. and Takamura, Y. (1985) Nitrogen and phosphorus runoff from watersheds. Tokyo University Press, Tokyo, Japan. 226 pp.

Takeuchi, M. (1992) An evaluation method of effects of self-purification of nitrogen in water. In:

Guides of evaluation methods of multi-functionality. (ed. by AFFRC and NIAES), AFFRC, Tokyo, Japan, and NIAES, Tsukuba, Japan. pp. 1-30.

Ukita, M. and Nakanishi, H. (1989) 1-2. Pollutant load developments from point sources. In: Modeling analysis of pollution in rivers. (ed. by Japanese Society of Civil Engineers), Gihodo shuppan, Tokyo, Japan. pp. 11-24.

Distribution and long-term changes in levels of polycyclic aromatic hydrocarbons in South Korean soils

Jae-Jak Nam[a], Back-Kyun Park[a], Kyu-Ho Sho[a] and Chang-Young Park[b]

[a] *National Institute of Agricultural Science and Technology, Suwon 441-707, Korea*
[b] *National Yeongnam Agricultural Experiment Station, Milyang 627-130, Korea*

Abstract

Polycyclic aromatic hydrocarbons (PAHs) in agricultural soils were measured using gas chromatography combined with ion trap mass spectrometry. To describe the national distribution of PAHs in South Korea, we collected 226 soil samples from paddy and upland soils, and recorded their locations with a GPS receiver. The distribution of PAHs was closely related to the pollution source, city size, and type of industry. The total PAH concentration in all samples averaged 236 $\mu g\ kg^{-1}$, and ranged from 23.3 to 2834 $\mu g\ kg^{-1}$. The highest concentrations were measured in soils close to iron-processing plants. The PAH content was fluoroanthene > benzo(b)fluoroanthene > pyrene. We also investigated the impact of industrialization on the agricultural environment over the past 20 years by analyzing soil collected from the Young-Nam Agricultural Experiment Station in southeastern Korea. The total PAH concentration had reached 200 $\mu g\ kg^{-1}$ by the early 1980s, and has not changed much since then. We conclude that PAHs have relatively little influence on the overall agricultural environment, but that some regions near pollution sources are threatened by these contaminants. We also verified that vertical movement of PAHs in soil occurs without artificial disturbances such as farming.

Keywords : PAHs, soil, GC-ITMS, GIS, monitoring, long-term change

1. Introduction

Polycyclic aromatic hydrocarbons (PAHs) originate mainly from anthropogenic sources. They are formed as byproducts of incomplete combustion of organic materials. The most significant anthropogenic sources of PAHs include heat and power generation from coal and other fossil fuels, coal production, petroleum refining, coal and oil shale conversion, and chemical manufacturing (Suess, 1976). Some PAHs are also derived from biogenic precursors such as pigments and steroids (Wakeham et al., 1980). In soils, PAHs can arise from a number of sources. Point sources include hydrocarbon spillage (Benner et al., 1990), incomplete combustion of fossil fuels (e.g., wood burning; Freeman and Catteil, 1990), the use of organic waste as compost and fertilizer (Smith et al., 2001), and power plants and blast furnaces (Van Brummelen et al., 1996). However, the majority of PAH pollution is likely to be caused by diffuse sources such as atmospheric deposition, and there is evidence that PAHs are transported over long distances by atmospheric movements (Aamot et al., 1996; Bakker et al., 2001; Halsall et al., 2001; Lunde and Bjorseth, 1977).

The deposited PAHs accumulate mainly in the soil's humus layer. The pathways of PAH dissipation in contaminated soil include volatilization, irreversible sorption, leaching, bioaccumulation by plants, and biodegradation (Reilly et al., 1996). PAH compounds with three or more rings tend to be strongly adsorbed to the soil. Strong sorption, very low water solubility, and very low vapor pressures make leaching and volatilization of PAHs insignificant pathways of PAH dissipation (Park et al., 1990). PAH concentrations in soil correlate significantly with the corresponding levels in the ambient air (Vogt et al., 1987); therefore, the determination of PAH levels in soils may provide important information on the levels of environmental pollution. The characteristic ratio of various PAHs and PAH profiles can both be used in qualitative and quantitative determination of the sources of PAHs. For example, the ratios of phenanthrene to anthracene and of fluoroanthene to pyrene have been commonly used as a means of determining the origins of PAHs (Gschwend and Hites, 1981; Vogt et al., 1987; Yang et al., 1991).

In the early 1970s, South Korea started its industrial revolution. The country's main industries are heavy industries such as iron processing, petroleum refining, and vehicle manufacturing. These industries may have adversely affected soil ecology and crop safety, but no investigations have yet been conducted in South Korea to address these issues. The concentrations and distributions of PAHs in South Korea have been studied mainly in sediments (Khim et al., 2001; Koh et al., 2002). In the present study, 16 parent PAH compounds included in the U.S. Environmental Protection Agency's priority pollutant list were selected because they are considered important in the assessment of the impact of industrialization on public health and the environment. Our study aimed to determine the concentrations of pollutant compounds in agricultural soils to map PAH pollution by using a geographical information system (GIS) and identify the probable sources of these compounds.

2. Materials and methods

2.1. Regional site description

For the purposes of sample collection, we divided the Korean peninsula into six categories (Fig. 1 and Table 1). These were chosen on the basis of geomorphic boundaries and to estimate the contribution of point and diffuse sources to PAH concentrations in the soil.

In the Korean peninsula, the prevailing wind comes from the northwest most of the year, but in summer, it switches to a predominantly southeasterly direction. The Seoul (SE) metropolitan area has about 20 million people and 6 million motor vehicles, as well as several

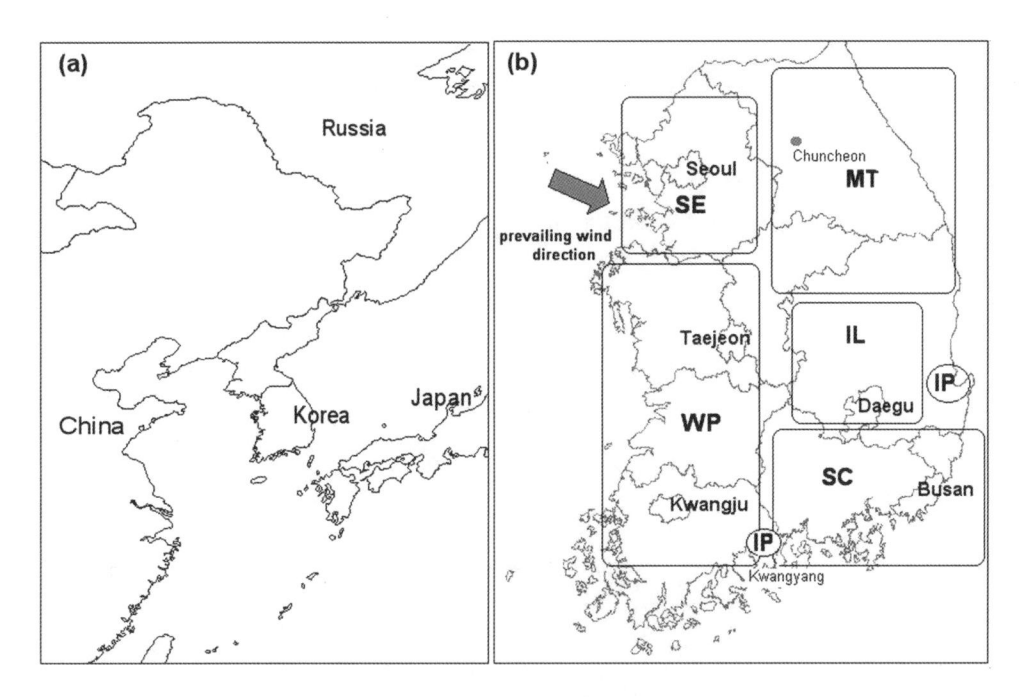

Fig. 1 Location of Korean Peninsula (a) and regional grouping of sampled sites (b) : SE, Seoul metropolitan area; MT, Mountain area ; IL, Inland area; SC, South coastal area ; WP, West plain area ; IP, Iron processing plant area

Table 1 Description of sectors from which soil samples were obtained

	Sector	Major city	Industrial activity	Pollutant source
SE	Seoul	Seoul	Complex	Industry, vehicles, coke burning
MT	Mountain	Chuncheon	Sightseeing	None
IL	Inland	Daegu	Textile industry	Industry, vehicles
SC	South-coastal area	Busan	Heavy industry	Industry, vehicles
WP	West Plain	Kwangju	Crop growing	Small industry
IP	Iron processing	Pohang Kwangyang	Iron processing	Coke burning, petrol refining

industrial plants. The atmospheric pollution load is high, and because of the prevailing wind, pollutants may be transported from Seoul into the mountain (MT) area to the east. The MT area has no local sources of PAH pollution; thus, any PAHs in the soils of this area have originated elsewhere and the area can serve as the basis for estimating long-range atmospheric transport of PAHs.

The western plain (WP) is mostly an agricultural area with few cities and no obvious local

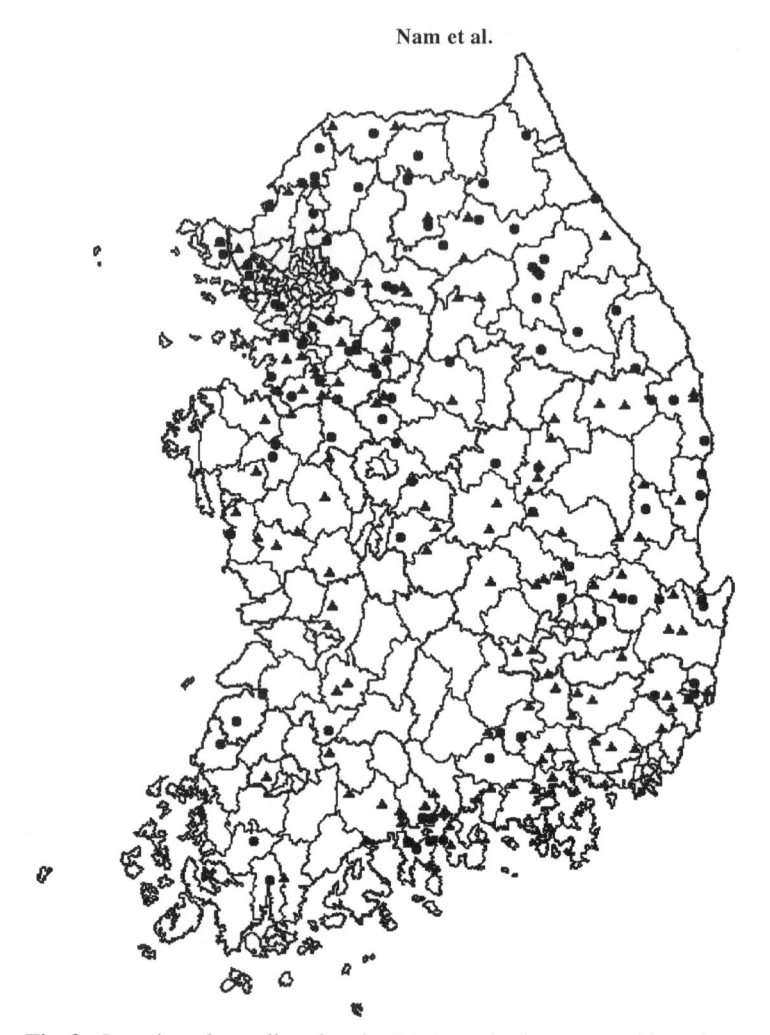

Fig. 2 Location of sampling sites for PAH monitoring : ○ , paddy soil; △ , upland soil

sources of PAH pollution. The inland (IL) area is bounded on the north, west, and east by mountain ranges and has a population of about 4 million. The area has a considerable amount of heavy industry that generates pollution, but may also receive diffuse pollution from the SE area. The south-coastal (SC) area is heavily populated (about 5 million) and is highly industrialized. Movement of PAH pollution into marine rather than terrestrial areas would be expected, except in summer. The final category of area that we defined (IP) has iron- and steel-processing plants, and was included to let us estimate the effect of heavy pollution on soil PAH concentrations.

2.2. Sample collection

Surface soils were collected in agricultural areas throughout South Korea at the locations shown in Figure 2. A total of 226 soil samples (126 paddy soils and 100 upland soils) were collected over 2 years to monitor the distribution of PAHs. A number of samples were collected from each location and bulked together to obtain a representative sample for analysis. To eliminate the direct effects of vehicle exhaust emissions, we collected samples at least 30 m away from the nearest road. Some sampling sites were selected specifically to cover specified contamination sources such as power plants and industrial enterprises.

The samples used to detect long-term changes in PAH levels were soils stored at the National Yeongnam Agricultural Experiment Station, Milyang. These soils were collected from the station's field to study the long-term application of fertilizer in paddy lands every year after harvesting. Since the late 1970s, the field has received applications of a sort of fertilizers and compost, and included a control plot. Samples were collected from both the control plot and the compost plot. The collected samples were dried at room temperature, sieved through a 20-mesh sieve and stored in glass bottles at 4 ℃.

2.3. PAH analysis

Samples (20 g) of air-dried soil were extracted for 16 h with dichloromethane (200 mL) in a Soxhlet apparatus. A silica gel column was used to purify the extract according to EPA method 3630 (EPA, 1994). The eluant was evaporated to about 1 mL prior to analysis. Deuterated PAHs (naphthalene-d_8, acenaphthene-d_{10}, phenanthrene-d_{10}, chrysene-d_{12}, perylene-d_{12}) were used as internal standards and were added to the soil prior to extraction. The standard curve was obtained by using 20, 50, 100, 500, and 1000 ng mL^{-1} of each PAH standard. All quantification was performed using internal standardization of the five internal standards at the 200 ng mL^{-1} level. The response curves for the 16 PAH compounds we analyzed were linear, and the correlation coefficients were all greater than 0.99. The moisture content was determined by heating the air-dried soil at 105 ℃ for 4 h to allow us to present all data on a dry-matter basis.

Quantitative analyses were performed on a gas chromatograph combined with an ion trap mass spectrometer (GCQ, ThermoFinnigan, USA) to determine levels of the following 16 PAHs: naphthalene, acenaphthylene, acenaphthene, fluorene, phenanthrene, anthracene, fluoranthene, pyrene, benz(a)anthracene, chrysene, benzo(b)fluoranthene, benzo(k)fluoranthene, benzo(a)pyrene, indeno-(1,2,3-cd)-pyrene, dibenzo(a,h)anthracene, and benzo(ghi)perylene. The mass spectrometer was operated in full-scan mode (mass range of 100 to 255 m/z) with electron-impact ionization at 70 eV. A fused silica capillary column (DB-5ms, 30 m × 0.25 mm I.D. × 0.25 μm film thickness; J&W Scientific Inc., USA) was used with helium as the carrier gas at a constant velocity of 40 cm s^{-1}. The temperatures of the injector, transfer line, and ion source were 285, 290, and 200 ℃, respectively. The oven temperature program started at 75 ℃, held for 5 min, increased to 150 ℃ at a rate of 25 ℃ min^{-1}, then increased to 265 ℃ at a rate of 4 ℃ min^{-1}. The latter temperature was held for 10 min, then increased to 285 ℃ at 30 ℃ min^{-1}. A 1-μl sample was injected in the splitless mode.

2.4. GIS mapping

The ArcView ver. 3.0 (ESRI, USA) geographic information system (GIS) was used to produce a distribution map of PAH concentrations, with concentrations displayed as a contour plot on the map. The position of each sample location was recorded using a Geo Explorer II

Table 2 Abbreviations, retention times, quantitation mass and MDL of PAH

Name	Abbreviation	Retention time (min)	Quantitation mass	MDL^3 (pg)
Naphthalene	Nap	5.62	128	7.14
Acenaphthylene	Acy	8.51	152	1.99
Acenaphthene	Ace	8.97	154	2.35
Fluorene	Fle	10.57	166	3.54
Phenanthrene	Phe	14.38	178	2.34
Anthracene	Ant	14.61	178	2.59
Fluoranthene	Fla	20.34	202	6.00
Pyrene	Pyr	21.49	202	3.06
Benzo (a) anthracene	BaA	28.40	228	3.89
Chrysene	Chr	28.58	228	2.90
Benzo (b) fluoranthene	BbF	34.16	252	2.65
Benzo (k) fluoranthene	BkF	34.28	252	6.05
Benzo (a) pyrene	BaP	35.28	252	2.81
Indeno (1, 2, 3-ced) pyrene	IcP	40.24	276	2.13
Dibenzo (a, h) anthracene	DaA	40.54	278	1.66
Benzo (g, h, i) perylene	BgP	41.59	276	2.86

[a] Minimum detection limit equals three times the standard deviation of the replicate analysis of the lowest standard (n=8).

Table 3 Concentrations and recoveries of PAH in NIST SRM 1941a using a Soxhlet extraction method (μg kg^{-1}, dry weight basis)

Name	Certified value^a	Soxhlet value^b	Percent of certified value	Soxhlet RSD^c
Naphthalene	1010 ± 140	792 ± 47	78.4	37.4
Acenaphthylene	(41 ± 10)^b	47 ± 4	115	3.55
Acenaphthene	(37 ± 14)	32 ± 7.7	87	6.80
Fluorene	97.3 ± 8.6	90 ± 11	93	9.60
Phenanthrene	489 ± 23	436 ± 34	89	29.7
Anthracene	184 ± 14	151 ± 13	82	11.2
Fluoranthene	981 ± 78	892 ± 25	91	21.8
Pyrene	811 ± 24	674 ± 31	83	27.6
Benzo (a) anthracene	427 ± 25	364 ± 31	85	27.4
Chrysene	380 ± 24	579 ± 29	152	25.3
Benzo (b) fluoranthene	740 ± 110	877 ± 26	119	22.9
Benzo (k) fluoranthene	361 ± 18	321 ± 6	89	5.26
Benzo (a) pyrene	628 ± 52	475 ± 20	76	17.3
Indeno (1, 2, 3-ced) pyrene	501 ± 72	453 ± 33	90	30.1
Dibenzo (a, h) anthracene	73.9 ± 9.7	133 ± 6.5	180	5.77
Benzo (g, h, i) perylene	525 ± 67	475 ± 15	90	13.5

[a] The uncertainty is based on a 95% confidence interval for the true concentration.
[b] Values in parenthesis were not certified.
[c] Relative standard deviations for the measured values are based on three Soxhlet replicate analysis.

GPS receiver (Trimble, USA). The WG84 coordinate system produced by the receiver was transformed into the UTM coordinate system to fit the actual map.

2.5. Quality control

The minimum detection level (MDL) ranged from 1.66 to 7.14 pg per individual PAH. The average recoveries of the PAHs ranged from 86% to 123% for sea sand spiked with 100 μg kg^{-1} of the PAH.

The abbreviations for each PAH that we analyzed, the retention times for each PAH, the ions used to obtain quantitative data, and the MDLs are shown in Table 2. Each MDL was based on three times the observed standard deviation of eight replicate analyses of the lowest standard. The limit of quantification was considered to be three times the MDL (Knoll, 1985; Mazzera et al., 1999; Long and Winefordner, 1983).

The method used in this study was verified using NIST SRM 1941a as a reference. The data are shown in Table 3. Recoveries of individual PAH compounds ranged from 76% to 180%.

3. Results and discussion

3.1. PAH concentrations in soil

The mean, median, and range of the PAH concentrations for individual PAHs and the total PAH concentration for agricultural soils are given in Table 4. The total PAH concentration varied by more than two orders of magnitude, ranging from 23.3 to 2834 μg kg^{-1}. The mean

Table 4 Mean, median, and range of PAH in Korean soils (μg kg^{-1}, dry weight basis)

PAH	Mean	Median	Range
Naphthalene	24.0	20.0	4.8 − 157
Acenaphthylene	2.1	0.56	<0.3 − 41.5
Acenaphthene	1.5	0.29	<0.35 − 33.7
Fluorene	3.6	1.9	<0.53 − 39.4
Phenanthrene	20.5	14.7	0.70 − 141
Anthracene	7.5	4.3	<0.30 − 43.1
Fluoranthene	33.7	21.7	2.6 − 353
Pyrene	26.0	14.7	1.6 − 317
Benzo (a) anthracene	18.6	10.5	<1.58 − 284
Chrysene	14.6	7.4	<0.44 − 267
Benzo (b) fluoranthene	31.8	19.9	<0.40 − 431
Benzo (k) fluoranthene	7.7	1.8	<0.91 − 138
Benzo (a) pyrene	16.3	9.5	<0.42 − 294
Indeno (1, 2, 3−ced) pyrene	12.2	2.9	<0.32 − 248
Dibenzo (a, h) anthracene	3.2	0.20	<0.25 − 120
Benzo (g, h, i) perylene	12.4	5.5	<0.43 − 221
Total PAH	236	158	23.3 − 2,834

N=226.

Table 5 Typical values for PAH in Korean paddy and upland soils in Korea (μg kg^{-1}, dry weight basis)

PAH	Paddy soil[a]		Upland soil[b]	
	Mean	Range	Mean	Range
Naphthalene	25.3	4.8 − 157	22.4	4.9 − 78.5
Acenaphthylene	2.2	<0.3 − 38.6	1.97	<0.43 − 41.5
Acenaphthene	1.5	<0.35 − 33.0	1.53	<0.35 − 33.7
Fluorene	4.6	<0.53 − 39.4	2.35	<0.53 − 37.6
Phenanthrene	22.4	0.70 − 120	18.0	3.3 − 141
Anthracene	8.0	0.30 − 43.1	6.88	<0.30 − 33.7
Fluoranthene	32.9	3.4 − 332	34.7	2.6 − 353
Pyrene	20.1	3.2 − 128	33.4	1.6 − 317
Benzo (a) anthracene	17.0	<1.58 − 75.8	20.5	<1.58 − 284
Chrysene	10.7	<0.44 − 50.9	19.4	<0.44 − 267
Benzo (b) fluoranthene	27.0	<0.40 − 201	37.8	<0.40 − 431
Benzo (k) fluoranthene	5.7	<0.91 − 50.1	10.1	<0.91 − 138
Benzo (a) pyrene	12.4	<0.42 − 70.1	21.1	<0.42 − 294
Indeno (1, 2, 3-ced) pyrene	7.7	<0.32 − 84.6	17.9	<0.32 − 248
Dibenzo (a, h) anthracene	2.3	<0.25 − 29.6	4.35	<0.25 − 120
Benzo (g, h, i) perylene	8.7	<0.43 − 78.6	17.0	<0.43 − 221
Total PAH	209	38.3 − 1,057	270	23.3 − 2,834

[a] N=226. [b] N=100.

total PAH concentration for all the samples we collected was 236 μg kg^{-1}, and the median was 158 μg kg^{-1}. These values are greatly in excess of the reported natural concentrations of PAHs in soil (1 to 10 μg kg^{-1}; Edward, 1987), but are slightly lower than the levels observed in rural soils in the United Kingdom, where the median value was 187 μg kg^{-1} (Wild and Jones, 1995). Therefore, typical South Korean soils from agricultural areas contained PAH concentrations that were greater than natural levels, but similar to those in soils from other highly industrialized countries (Jones et al., 1989; Trapido, 1999).

Table 5 summarizes the mean concentration and range of concentration values for PAHs in upland and paddy soils. The concentration in upland soils (mean: 270 μg kg^{-1}) was about 35% higher than that in paddy soils (mean: 209 μg kg^{-1}). This difference may be accounted for by the fact that paddy soils are irrigated and flooded for about 6 months each year, and then are drained after harvesting, and thus PAHs may be leached from these soils.

The highest concentrations of PAHs were found in soils from areas adjacent to iron-processing plants that burn large amounts of coal. The heavy contamination was restricted to a small area because the plants were located in coastal areas, where a large proportion of the contaminants would be exported to the marine environment under the prevailing northwest winds. However, contamination of agricultural soils is most likely to be caused by the long-range atmospheric movement of PAHs from point sources of pollution to remote rural sites (Aamot et al., 1996; Lunde and Bjorseth, 1977; Halsall et al., 2001).

Table 6 Characterization of PAH for Each Area[a] (μg kg^{-1}, dry weight basis)

PAH	SE	MT	IL	SC	WP	IP
No. of sample	61	50	24	31	45	15
Total PAH	257 (232)[b]	271 (317)	199 (154)	171 (76)	118 (66)	578 (704)
Median	185	160	165	160	108	321
Range of concentration	32 − 1057	43 − 1,852	43 − 623	41 − 336	23 − 303	105 − 2833
2-4 rings PAH	169 (198)	167 (204)	122 (95)	116 (64)	91 (70)	338 (378)
5-6 rings PAH	87 (115)	103 (149)	77 (107)	55 (42)	28 (29)	240 (363)

[a] SE: Seoul metropolitan area; MT: mountain area; IL: Inland area; SC: south coastal area; WP: west plain area; IP: iron processing plant area.

[b] Numbers in parentheses are standard deviations.

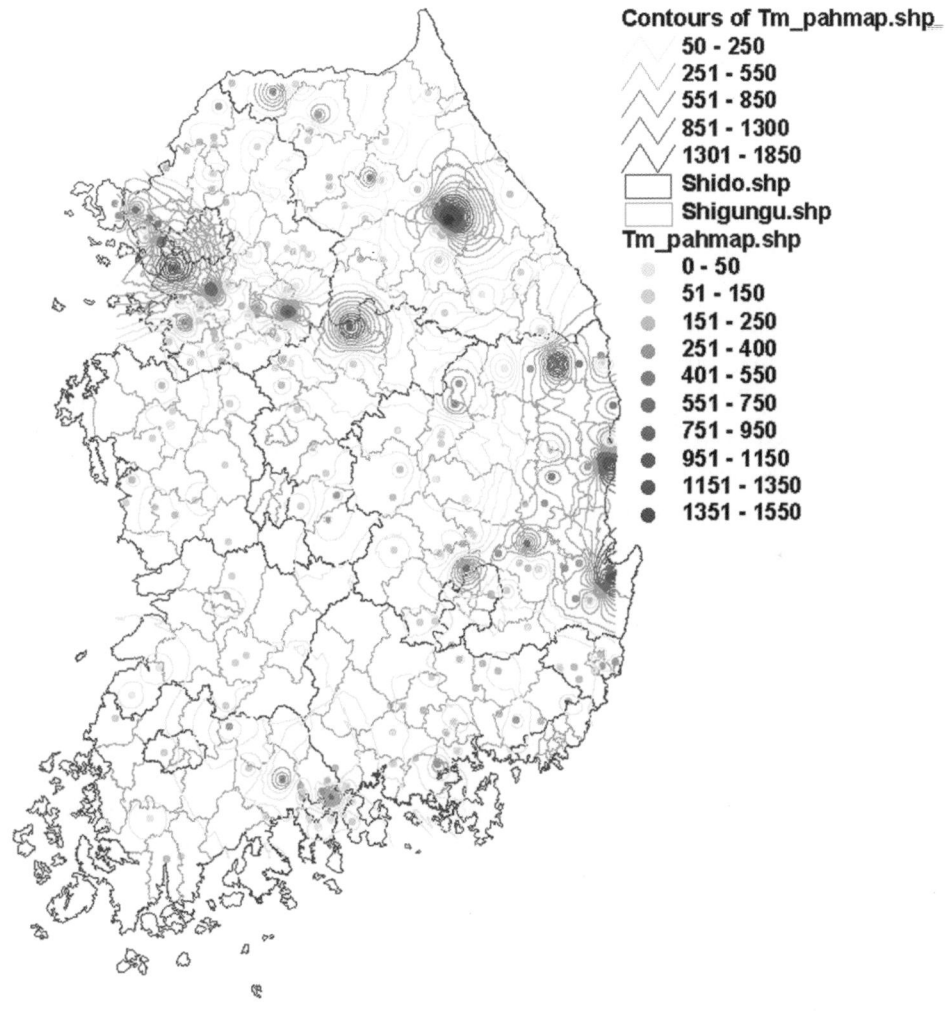

Fig. 3 Distribution map of total PAH in Korea

3.2. Distribution of PAHs

The spatial distribution of total PAH concentrations in relation to geomorphic locations is shown in Table 6 and Figure 3. For each area, PAH concentrations averaged 257 (SE), 271 (MT), 199 (IL), 171 (SC), 118 (WP), and 578 (IP) μg kg^{-1}, respectively.

The PAH concentrations in soils from the mountain (MT) area ranged from 43 to 1852 μg kg^{-1} and averaged 271 μg kg^{-1}, whereas those from the Seoul (SE) area ranged from 32 to 1057 μg kg^{-1} and averaged 257 μg kg^{-1}. There were no point sources of pollution in the MT area, and because of the prevailing wind direction (from the northwest), it can be assumed that the primary origin of the PAHs was from the SE area, an area that generates a significant amount of pollution because of its large number of vehicles and its industries. The soils in the MT area act as a sink for PAH pollutants originating in the west of the country, but because this region's mountains act as a rain barrier, PAHs are deposited primarily through wet or dry deposition (Park et al., 2000).

The average PAH concentrations were similar in the IL and SC areas. The PAH concentration ranged from 43 to 623 μg kg^{-1} (mean: 199 μg kg^{-1}) in the IL area, and from 41 to 336 μg kg^{-1} (mean: 171 μg kg^{-1}) in the SC area. The slightly higher soil PAH concentrations in the IL area are due to the fact that this area is surrounded on three sides (west, north, and east) by mountains that can block and control the flow of wind, and thus receives less PAH input from wind than in other areas, whereas the prevailing northwest wind disperses pollution in a seaward direction in the SC area, which is heavily industrialized and which has heavy vehicular traffic.

The concentration of PAHs in the soils of the WP area, which is the country's main rice production area, ranged from 23 to 303 μg kg^{-1} and averaged 118 μg kg^{-1}. There are few pollution sources in this area, and the origin of PAHs in the area's soils may be mainland Asia as well as sources within South Korea itself.

In comparison with the other areas, the two IP areas exhibited high soil PAH concentrations. The maximum value obtained was 2833 μg kg^{-1}, with a mean of 578 μg kg^{-1}. The high concentration of PAHs in these soils suggest that emissions from iron-processing plants in these areas, which are among the largest in the world, contain high levels of PAHs that are deposited on the surrounding soil. Because these plants are situated on the coast, the area affected by their pollution is fairly small and discrete.

From the data presented in this paper, it is evident that soils near the pollution source are generally the most heavily contaminated, although the soils in the MT area had higher PAH levels than those from the SE area. This observation can be accounted for by the influence of the prevailing winds and geomorphic factors.

3.3. PAH profiles

The profiles of each PAH present in paddy and upland soils are presented in Table 5. The general profile of PAHs was similar in both soils, with the predominant PAH being fluoranthene (Fla), benzo(b)fluoranthene (BbF), and pyrene (Pyr), which contain five, five, and four rings, respectively. Concentrations of PAHs were generally higher in upland soils, and this

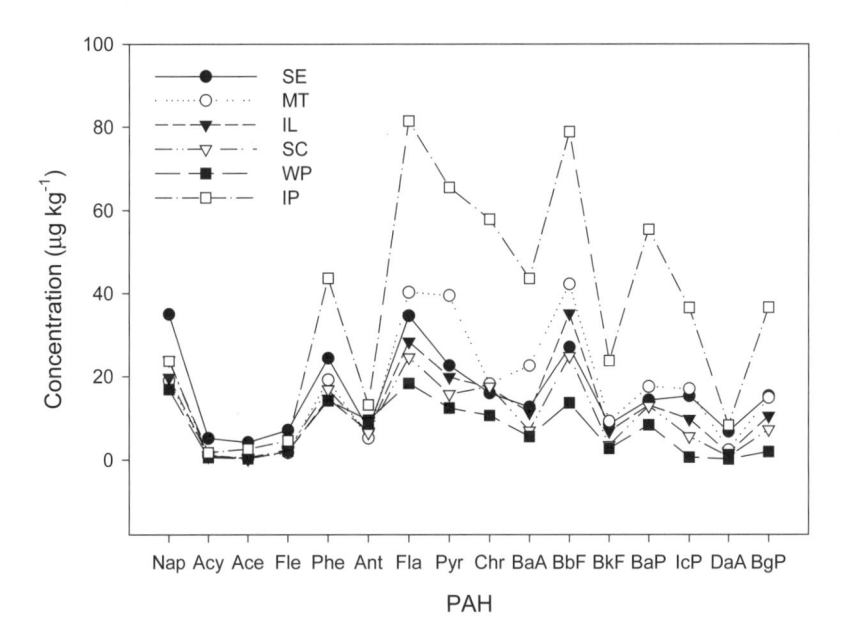

Fig. 4 Typical PAHs profile for each : SE, Seoul metropolitan area ; MT, Mountain area ; IL, Inland area ; SC, South coastal area ; WP, West plain area ; IP, Iron processing plant area

may be due to a number of factors; these include the fact that the upland soils contain a higher organic matter content than paddy soils (PAHs will associate with organic soils to a greater extent than with mineral soils) and the fact that paddy soils are saturated with water for a significant part of the year, and thus there is an increased likelihood of leaching of PAHs into the groundwater. It should also be noted that the spatial distribution of PAHs can vary considerably within a sampling area.

The distributions of individual PAH compounds throughout the study areas were similar, except in areas MT and IP (Fig. 4). In the MT area, the concentrations of pyrene and benzo(a)anthracene were significantly higher than in all other areas except IP. Although the major source of PAHs in the MT area is likely to be from the SE area due to the prevailing wind direction, it is evident that because the PAH profiles for MT and SE do not match, PAHs are being transported into the MT area from other areas as well, possibly including mainland Asia. However, further investigations would be required to prove this hypothesis.

The distribution of PAHs with two or three rings was similar in all areas, but it is noticeable that the concentrations of PAHs with four to six rings were significantly higher in the MT area than in other areas. In particular, benzo(a)pyrene was present in high concentrations, and although the mechanism responsible for the formation of this highly toxic compound is not known, there is little doubt that its source was in the emissions from the iron -processing plants.

The ratios between various pairs of individual PAH compounds have often been used as a method of identifying the most significant sources of PAHs in a given sample. For example, Yang et al. (1991) reported that a ratio of about 3 for Phe/Ant indicated PAHs arising primarily from motor vehicle exhaust, whereas a ratio greater than 50 indicated that the major source was the combustion of mineral oil. Similarly, a ratio of 1 for Fla/Pyr indicated that the PAH origin was likely to be from combustion processes, whereas a ratio greater than 1 suggested that the origin was petroleum refining. On the basis of these observations and the data presented in Tables 4 and 5, it appears that pyrogenic origins such as motor vehicle exhaust and heavy industry emissions are the dominant sources of PAHs in South Korean soils.

3.4. Long-term changes in PAHs

The changes in the total PAH concentration during the last two decades are presented in Figure 5. The figure shows that before the 1980s, the total PAH concentration was already greater than the 209 $\mu g\ kg^{-1}$ average concentration for paddy soils throughout the country. Although we expected that PAH concentrations would increase over time in the country's soils, the concentrations have actually decreased slightly with some fluctuations. These results are mainly due to the fact that the sample site for the control was located in a rural area that was largely free from contamination caused by industrialization. Another possibility is that the PAH concentration in the rural area was reached at the balance to some extent around this concentration.

PAH concentrations as a function of each compound's molecular weight (MW) are shown in Figure 6. The heavier PAHs (MW >252) showed almost no change over the past 20 years, but PAHs with medium MW (MW from 178 to 228) decreased slightly during this period. We believe that the decrease in the medium-MW PAHs affected the overall decreases in total PAH concentration during this period. The lighter PAHs (MW <178) also remained largely

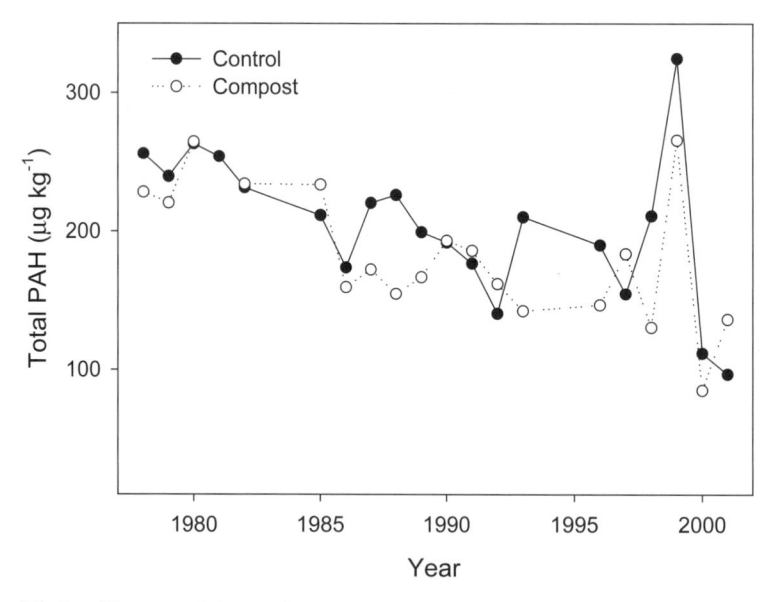

Fig 5. Changes of the total PAH concentration in soils for the last 20 years

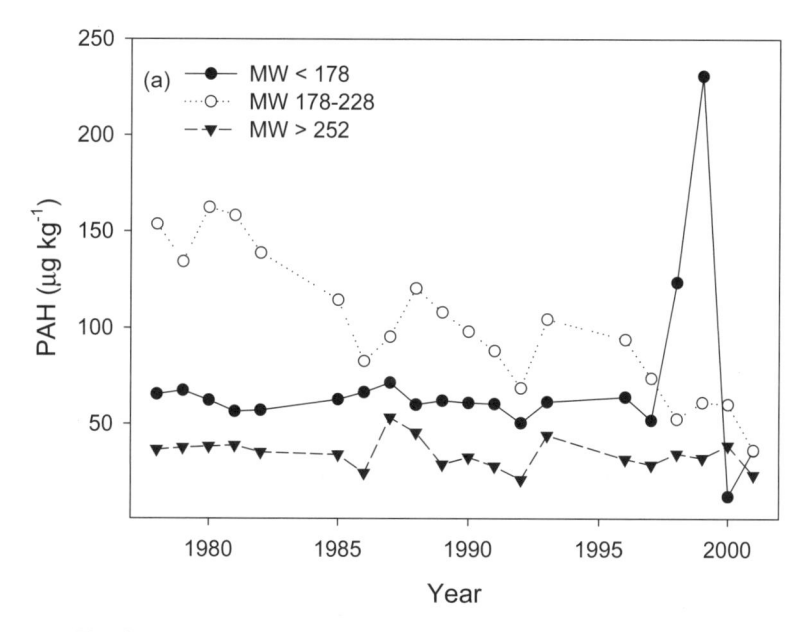

Fig. 6 Changes of the PAH concentrations in soils by a sort of PAHs

constant despite a large increase in 1999. However, we could find no reason for this increase, and it didn't affect the next year's results because of the properties of naphthalene, which is a little volatile and water miscible.

We believe that the overall decrease in PAH concentrations in the soil despite increasing numbers of vehicles and increasing industrialization arises from declining use of coal since the late 1980s. Consequently, the agricultural environment, and particularly soils, may be reaching an equilibrium in terms of these organic pollutants. However, we did not study the direct contamination of crops by direct deposition of particulates, including the pollutants. We plan to concentrate our future research in this area.

References

Aamot, E., Steinnes, E. and Schmid, R. (1996) Polycyclic aromatic hydrocarbons in Norwegian forest soils: Impact of long range atmospheric transport. *Environ. Pollut*. 92: 275 – 280

Bakker, M.I., Casado, B., Koerselman, J.W., Tolls, J. and Kollöffe, C. (2001) Polycyclic aromatic hydrocarbons in soil and plant samples from the vicinity of an oil refinery. *Science of the Total Environ*. 263: 91 – 100

Benner, B.A. Jr, Bryner, N.P., Wise, S. A., Mulholland, G.W., Lao, R.C. and Fingas, M.F. (1990) Polycyclic aromatic hydrocarbon emission from the combustion of crude oil on water. *Environ. Sci. Technol*. 24: 1418 – 1427

Edward, N.T.J. (1987) Polycyclic aromatic hydrocarbons (PAHs) in the terrestrial environment - a review. *J. Environ. Qual*. 12: 427 – 441

EPA (1994) Test methods for evaluating solid waste, physical/chemical methods. Environmental Protection Agency, Office of Solid Waste and Emergency Response, Washington, D. C. SW-

846, Revision 2.

Freeman, D.J. and Catteil, F.C.R. (1990) Wood burning as source of atmospheric polycyclic aromatic hydrocarbons. *Environ. Sci. Technol.* 24: 1581-1585

Gschwend, P.M. and Hites, R.A (1981) Fluxes of the polycyclic aromatic hydrocarbons to marine and lacustrine sediments in the northeastern United States. *Geochimica et Cosmochimica Acta* 45: 2359-2367

Halsall, C.J., Sweetman, A.J., Barrie, L.A. and Jones, K.C. (2001) Modeling the behavior of PAHs during atmospheric transport from the UK to the Arctic. *Atmospher. Environ.* 35: 255-267

Jones, K.C., Stratford, J.A., Waterhouse, K.S., Furlong, E.T., Giger, W., Hites, R.A., Schaffner, C. and Johnston, A.E. (1989) Increases in the polynuclear aromatic hydro-carbon content of an agricultural soil over the last century. *Environ. Sci. Technol.* 23: 95-101

Khim, J.S., Lee, L.T,, Kannan, K., Villeneuve, D.L., Giesy, J.P. and Koh, C.H., (2001) Trace organic contaminants in sediment and water from Ulsan Bay and its vicinity, Korea. *Archives Environ. Contam. Toxicol.* 40: 141-150

Knoll, J.E. (1985) Estimation of the limit of detection in chromatography. *J. Chromatogr. Sci.* 23: 422-425

Koh, C.H., Khim, J.S., Lee, K.T., Villeneuve, D.L., Kannan, K. and Giesy, J.P. (2002) Analysis of trace organic contaminants in environmental samples from Onsan Bay, Korea. *Environ. Toxicol. Chem.* 21: 1796-1803

Long, G.L. and Winefordner, J.D. (1983) Limit of detection, A closer look at the IUPAC definition. *Anal.l Chem.* 55: 712A-715A

Lunde, G. and Bjorseth, A. (1977) Polycyclic aromatic hydrocarbons in long-range transported aerosol. *Nature* 268: 518-519

Mazzera, D.T.T., Hayes, T., Lowenthal, D. and Zielinska, B. (1999) Quantitation of polycyclic aromatic hydrocarbons in soil at McMurdo station, Antarctica, *Science of the Total Environ.* 229: 65-71

Park, K. S., Sims, R. S., Dupont, R. R., Doucette, W. J. and Mathews, J. E. (1990) Fate of polycyclic aromatic hydrocarbon compounds in two soil types: influence of volatilization, abiotic loss and biological activity. *Environ. Toxicol. Chem.* 9: 187-195

Park, S.U., In, H.J., Kim, S.W. and Lee, Y.H. (2000) Estimation of sulfur deposition in South Korea. *Atmospher. Environ.* 34: 3259-3269

Reilley, K.A., Banks, M.K. and Schwab, A.P. (1996) Dissipation of polycyclic aromatic hydrocarbons in the rhizosphere. *J. Environ. Qual.* 25: 212-219

Smith, K. E. C., Green, M., Thomas, G. O. and Jones, K. C. (2001) Behavior of sewage sludge-derived PAHs on Pasteur. *Environ. Sci. Technol.* 35: 2141-2150

Suess, M. J. (1976) The environmental load and cycle of polycyclic aromatic hydrocarbons. *Science of the Total Environ.* 6: 239-250

Trapido, M. (1999) Polycyclic aromatic hydrocarbons in Estonian soil: contamination and profiles. *Environ. Pollut.* 105: 67-74

Van Brummelen, T.C., Verweij, S.A., Wedzinga, S.A. and Van Gestel, C.A.M. (1996) Enrichment of polycyclic aromatic hydrocarbons in forest soils near a blast furnace plant. *Chemosphere* 32:

293–314

Vogt, N.B., Brakstad, F., Thrane, K., Nordenson, S., Krane, J., Aamot, E., Kolset, K., Esbensen, K. and Steinnes, E. (1987) Polycyclic aromatic hydrocarbons in soil and air: statistical analysis and classification by SIMCA method. *Environ. Sci. Technol.* 21: 35–44

Wakeham, S.G., Schaffner, C. and Giger, W. (1980) Polycyclic aromatic hydrocarbons in recent lake sediment-II. Compounds derived from biogenic precursors during early digenesis. *Geochimica et Cosmochimica Acta* 43: 27–33

Wild, S R. and Jones, K.C. (1995) Polynuclear aromatic hydrocarbons in the United Kingdom environment: A preliminary source inventory and budget. *Environ. Pollut.* 88: 91–108

Yang, S.Y.N., Connell, D.W., Hawker, D.W. and Kayal, S.I. (1991) Polycyclic aromatic hydrocarbons in air, soil, and vegetation in the vicinity of an urban roadway. *Science of the Total Environ.* 102: 229–240

Temporal changes in dioxin levels in Japanese paddy fields

Nobuyasu Seike and Takashi Otani

National Institute for Agro-Environmental Sciences, 3-1-3 Kannondai, Tsukuba, Ibaraki 305-8604, Japan

Abstract

Paddy soils collected from across Japan since 1960 were analyzed for PCDD/Fs and dioxin-like PCBs contained in PCP and CNP herbicides as impurities. PCDD/Fs derived from PCP differed greatly in concentration and composition between production methods (chlorination of phenol and hydroxylation of hexachlorobenzene). Total emissions of both types of compound were 0.84 to 1300 kg-TEQ (mean: 120 kg-TEQ) from PCP and 50 to 760 kg-TEQ (mean: 180 kg-TEQ) from CNP. Emission of both compounds from PCP increased in the early 1960s, then decreased rapidly in the 1970s. In contrast, emission from CNP increased throughout the late 1960s. Concentrations of PCDD/Fs in paddy soils increased from the 1960s through the 1970s, then decreased until the present. The production of PCP and CNP correlated well with changes in the concentrations of OCDD and 1,3,6,8-TeCDD in paddy soils. Before nationwide use of PCPs, the fallout from incinerators was the major source of PCDD/Fs in paddy soils, but PCDD/F concentrations and compositions in paddy soils changed significantly after nationwide use of PCP and CNP began in the 1960s. From the 1980s onward, negligible changes in PCDD/F compositions were observed because of reduced PCDD/F emissions from PCP and CNP. Instead, the proportion of PCDD/Fs from incinerators gradually increased in paddy soils over the last 20 years.

Keywords : Dioxins, temporal changes, paddy soils, PCP, CNP

1. Introduction

It is well known that polychlorinated dibenzo-*p*-dioxins and dibenzofurans (collectively, PCDD/Fs) are released into the environment through the incineration of rubbish and waste. A variety of chlorinated compounds such as polychlorinated biphenyls (PCBs; Wakimoto et al., 1988), pentachlorophenol (PCP; Hagenmainer and Brunner, 1987; Masunaga et al., 2001a) and 2,4,6-trichlorophenyl-4-nitrophenylether (chlornitrofen, CNP; Yamagishi et al., 1981; Masunaga et al., 2001a) contain PCDD/Fs as significant impurities. PCP and CNP have been commonly used as herbicides in paddy fields across Japan in the past. PCP was widely used through the 1960s and the early 1970s, whereas CNP started to be used in the late 1960s. It has been estimated that totals of roughly 170×10^6 kg of PCP and 78×10^6 kg of CNP (equivalent to the active ingredient) have been used in Japan. PCDD/F isomers such as OCDD and 1,3,6,8-/1,3,7,9-TeCDD are major impurities of PCP and CNP, respectively, and are widely distributed in aquatic environments (Seike et al., 1994, 1998; Masunaga et al., 2001b). It is most likely that the PCDD/Fs applied to paddy fields have since entered and polluted surrounding watersheds and aquatic environments. It is therefore imperative to trace the temporal variation in PCDD/Fs in paddy soils so as to elucidate the contamination pathways of the chemicals and their dynamics in agricultural environments.

Sediment cores (Pearson et al., 1997; MacDonald et al., 1998), vegetation (Kjeller et al., 1996), and soils (Kjeller et al., 1991; Alcok et al., 1998) are all very useful samples for tracing temporal changes in PCDD/F levels in the environment. Previous studies have shown that PCDD/F contamination began as long ago as 1900, but reached a maximum during the period between the 1960s and the 1980s. Although PCDD/F concentrations in the agricultural environment have progressively decreased in recent years, it remains important to analyze trends in PCDD/F concentrations and compositions in paddy soils so as to predict the half-lives of dioxins in these soils.

Paddy soils collected from across Japan since 1960 have been preserved in the soil inventory of the National Institute for Agro-Environmental Sciences (NIAES). In the present paper, the origins of PCDD/Fs and of dioxin-like PCBs are discussed in relation to the use of the herbicides PCP and CNP. We discuss the results of our analysis of paddy soils preserved since 1960 at NIAES to trace the changes in PCDD/F concentrations and compositions. We also evaluate the contributions of impurities in PCP and CNP to contamination by PCDD/Fs in paddy soils.

2. Materials and Methods

We analyzed PCDD/Fs and dioxin-like PCBs in 10 PCP and 24 CNP samples produced from 1966 to 1985 and from 1972 to 1994, respectively. Air-dried and sieved (<2 mm) paddy soils collected from 17 locations across Japan since 1960 were preserved in polyethylene bottles at room temperature. In the present study, PCDD/Fs and dioxin-like PCBs in samples from 5 locations in Japan were analyzed. Before extraction, we removed the surface layer of the soils stored in the polyethylene bottles. All paddy soil samples were extracted with toluene following the Soxhlet method. Purification and separation were carried out using multilayer silica gel, alumina, and activated carbon column chromatography. Samples were analyzed using a high-resolution gas chromatograph combined with a mass spectrometer (HRGC/HRMS; HP6890/Auto Spec-Ultima,) equipped with an SP-2331 column (SPELCO) and a DB-5MS column (J&W Scientific).

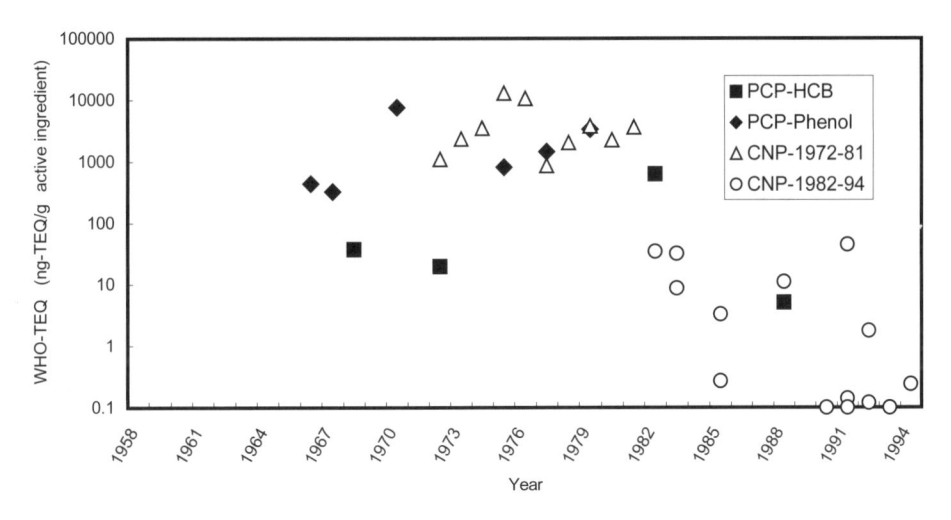

Fig. 1 Temporal changes of TEQ in PCP and CNP

3. Results and Discussion

3.1. Dioxin concentrations and compositions in PCP and CNP

The results of our analyses of the PCDD/Fs and dioxin-like PCBs in the 10 PCP and 24 CNP samples are shown in Figure 1. The TEQ (World Health Organization, 2,3,7,8-TeCDD toxicity equivalency quantity) values in PCP ranged from 4.5 to 7500 ng-TEQ g^{-1} (geometric mean: 310 ng-TEQ g^{-1}). Concentrations of PCDD/F isomers increased with an increasing degree of chlorination of the homologues. Most dioxin-like PCB isomers were present, if at all, at concentrations below the quantification limit (between <1 and <1000 ng g^{-1}, depending on the compound). The major procedures for synthesizing PCP are the hydroxylation of hexachlorobenzene (the HCB method) and the chlorination of phenol (the phenol method). TEQ values in PCP synthesized by the two methods ranged from 4.5 to 630 ng-TEQ g^{-1} (geometric mean: 38 ng-TEQ g^{-1}) and from 320 to 7500 ng-TEQ g^{-1} (geometric mean: 1300 ng-TEQ g^{-1}), respectively. We found higher concentrations of PCDD/Fs in PCP synthesized using the phenol method than in those synthesized using the HCB method.

The percentage contribution to TEQ by PCDD/F isomers is usually used to evaluate the environmental toxicity of the isomers. In this study, however, we used this percentage to evaluate the normalized 2,3,7,8-substituted isomer compositions because of the large differences in concentration we observed between low- and high-concentration isomers in PCP, CNP, and paddy soils. Figure 2 shows the percentage contribution to TEQ by 2,3,7,8-substituted isomers in PCP synthesized using the HCB and phenol methods. The highest percentage contribution was for 1,2,3,4,6,7,8-HpCDD in PCP synthesized using either method. Except for this isomer, percentage contributions decreased in the following order: OCDD > 1,2,3,4,7,8-HxCDD > 1,2,3,7,8-PeCDD > 1,2,3,4,6,7,8-HpCDF[GHT23] in PCP synthesized using the HCB method. In contrast, the order was 1,2,3,6,7,8-HxCDD >

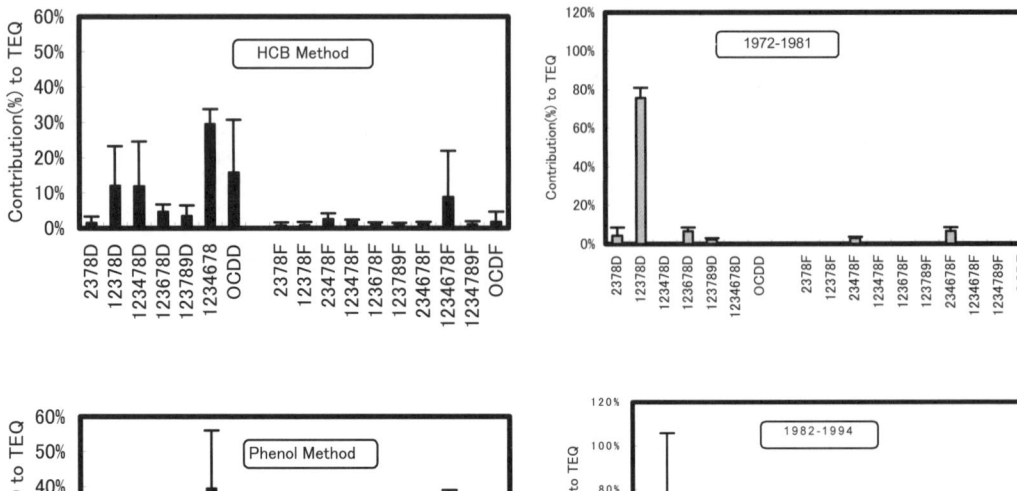

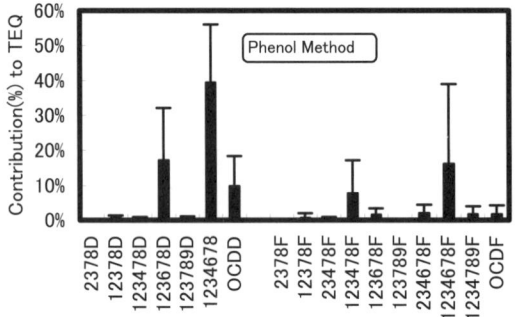

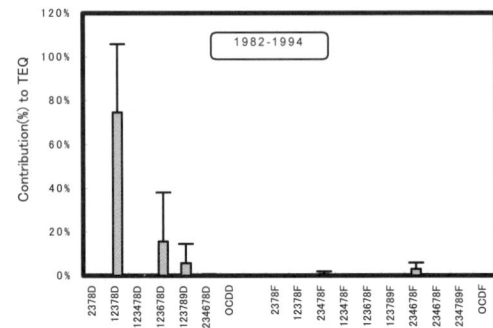

Fig. 2 Contribution(%) to TEQ by 2,3,7,8-substituted 2, 3, 7, 8-substituted isomers in PCP
Bars denote standard deviation of means

Fig. 3 Contribution(%) to TEQ by isomers in CNP
Bars denote standard deviation of means

1,2,3,4,6,7,8-HpCDF > OCDD > 1,2,3,4,7,8-HxCDF in PCP synthesized using the phenol method. These results suggest that the concentrations and compositions of PCDD/Fs in PCP differed in the different manufacturing methods.

TEQ values in CNP ranged from <0.1 to 13 000 ng-TEQ g^{-1} (geometric mean: 33 ng-TEQ g^{-1}). Isomer concentrations decreased with an increasing degree of chlorination of the homologues, such as TeCDD/Fs. TEQ values in CNP decreased drastically after 1981 (Fig. 1). From 1972 to 1981 and from 1982 to 1994, TEQ in CNP ranged from 860 to 13 000 ng-TEQ g^{-1} (geometric mean: 4300 ng-TEQ g^{-1}) and from <0.1 to 45 ng-TEQ g^{-1} (geometric mean: 1.3 ng-TEQ g^{-1}), respectively. Figure 3 shows the percentage contribution to TEQ by 2,3,7,8-substituted isomers in CNP synthesized before and after 1981. The greatest contributor to TEQ was 1,2,3,7,8-PeCDD. Few differences in the composition of 2,3,7,8-substituted isomers were found between CNP synthesized before and after 1981 (Fig. 3).

3.2. Estimation of dioxin emissions from PCP and CNP between 1958 and 1994

The dioxin emissions from PCP and CNP between 1958 and 1994 were estimated by multiplying TEQ values by the circulation of these pesticides in each year. Dioxin emissions from PCP and CNP were estimated to be 0.84 to 1300 kg-TEQ (mean: 120 kg-TEQ) and 50

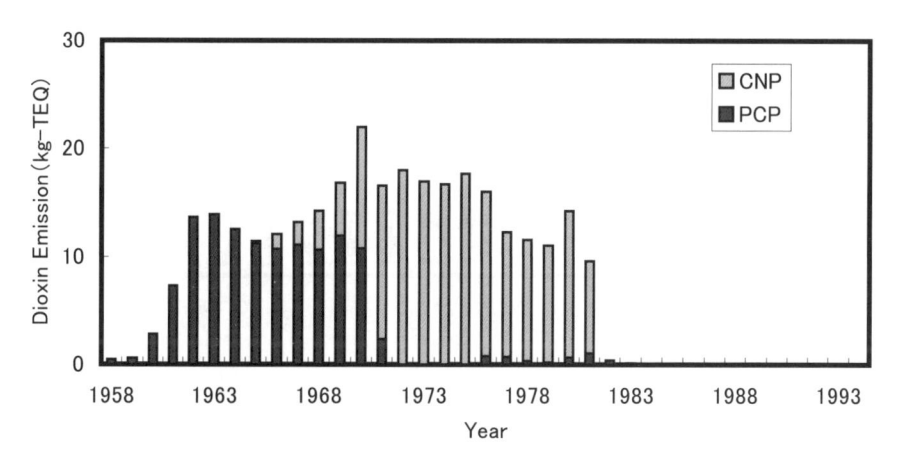

Fig. 4 Time trend of estimated dioxin emissions from PCP and CNP from 1958 to 1994

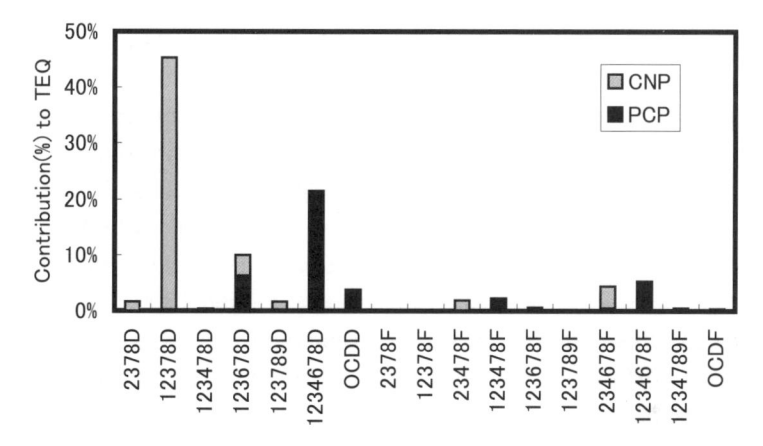

Fig. 5 Contribution (%) to TEQ by 2, 3, 7, 8-substituted congeners in PCP and CNP

to 760 kg-TEQ (mean: 180 kg-TEQ), respectively (Fig. 4). The dioxin emissions estimated by Masunaga et al. (2001a) were comparable to those in the present study: 440 kg-TEQ and 220 to 270 kg-TEQ from PCP and CNP, respectively. Figure 4 suggests that dioxin emissions from PCP were greatest during the 1960s. In contrast, emissions from CNP increased until the mid-1960s, decreased slowly until 1981, then decreased rapidly thereafter.

The emission of highly chlorinated isomers such as OCDD/F and of less-chlorinated isomers such as 2,3,7,8-TeCDD and 1,2,3,7,8-PeCDD from PCP and CNP were greatest from 1985 to 1994. Figure 5 shows the percentage contribution to TEQ by 2,3,7,8-substituted isomers. Contributions to TEQ by these isomers were generally low. The two major contributors to TEQ were 1,2,3,7,8-PeCDD from PCP and 1,2,3,4,6,7,8-HpCDD from CNP.

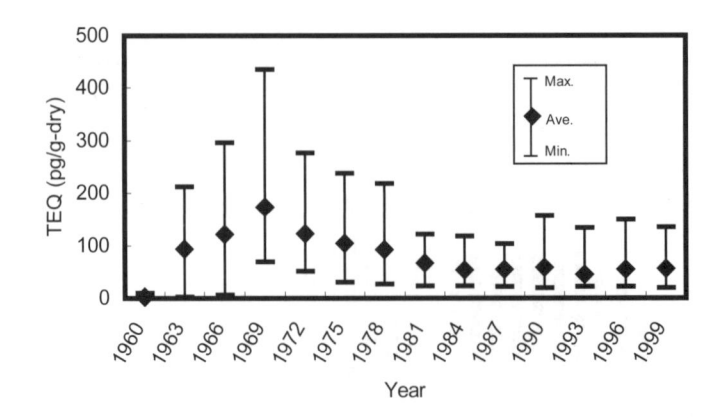

Fig. 6 Temporal change of TEQ in Japanese paddy soils (5 locations)

3.3 Temporal changes in dioxin concentrations in paddy soils

TEQ values measured in 1999 ranged from 20 to 130 pg-TEQ g^{-1} (mean: 55 pg-TEQ g^{-1}), and agreed with the results (5.3 to 180 pg-TEQ g^{-1}; mean: 44 pg-TEQ g^{-1}) of another study in Japan (Ministry of Environment and Ministry of Agriculture, Forestry and Fisheries, 2001). We therefore believe that the preserved samples represent typical paddy soils.

Temporal changes in TEQ in paddy soils from 1960 to 1999 are shown in Figure 6. TEQ values ranged from 1.7 to 440 pg-TEQ g^{-1} during this period and increased drastically during the 1960s, but began decreasing in the 1970s. Kjeller et al. (1991) analyzed PCDD/Fs in samples of cultivated soils preserved at the Rothmansted Experimental Station since the 1840s. Concentrations of PCDD/Fs started to increase around 1900. It was thought that the major source of PCDD/Fs in these soils was atmospheric deposition of compounds emitted by combustion processes. However, the Rothmansted TEQ in 1966 was much lower than that in the paddy soils in the present study. This suggests that there were significant inputs of dioxins from sources other than combustion processes in Japan during the 1960s and the 1970s.

Our data suggest a half-life of 10 years for dioxins. Other workers have estimated a half-life for dioxins in soil of between 6 and 20 years (Cerlesi et al., 1989; Maclachlan et al., 1996). Additional analyses of PCDD/Fs in more paddy soil samples will be required before we can calculate the isomer-specific half-life of various dioxins.

3.4. Temporal changes in dioxin sources in paddy soils

Figure 7 shows the temporal changes in PCP and CNP circulation (the amount of the active ingredient) in Japan and the corresponding OCDD and 1,3,6,8-/1,3,7,9-TeCDD concentrations in paddy soils in Japan. Both OCDD and 1,3,6,8-/1,3,7,9-TeCDD are major impurities in PCP and CNP. Temporal changes in the concentrations of these isomers were closely correlated with PCP and CNP circulation values. This suggests that the significant increase in PCDD/F concentrations and TEQ values were due to the increased use of PCP and CNP in paddy fields.

Figure 8 shows the percentage contribution to TEQ by 1,2,3,7,8-PeCDD, 1,2,3,6,7,8-

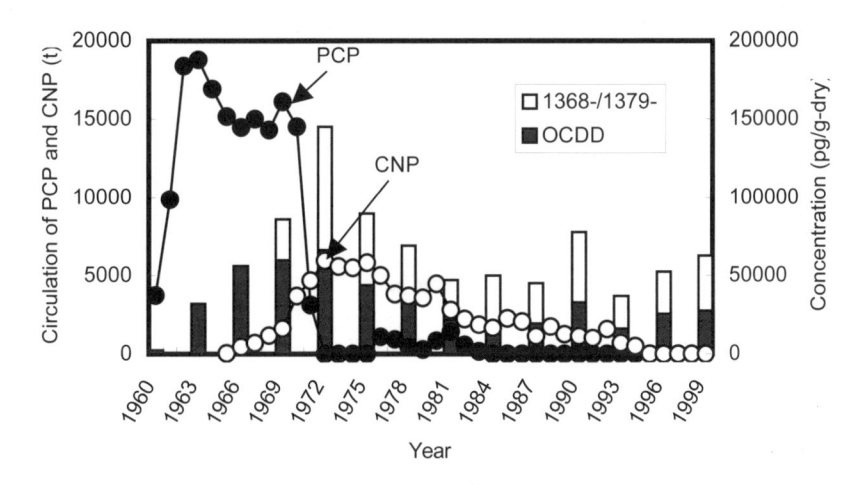

Fig. 7 Temporal changes of PCP and CNP circulation and OCDD, 1, 3, 6, 8- and 1,3,7,9-TeCDD concentrations in paddy soils

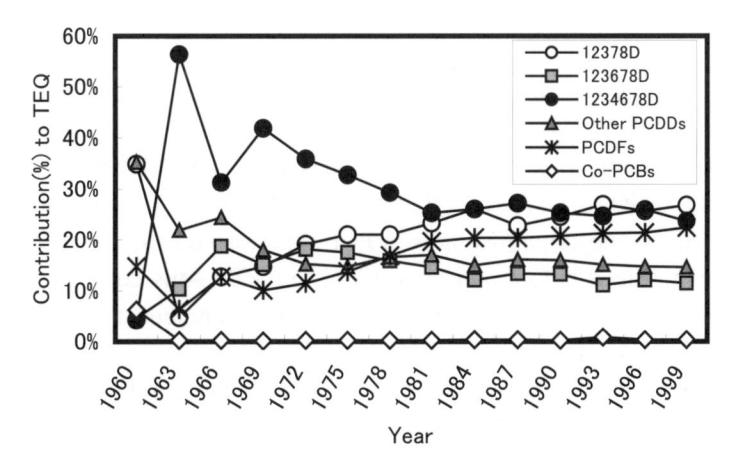

Fig. 8 Contribution (%) to TEQ by 1, 2, 3, 7, 8-PeCDD, 1, 2, 3, 6, 7, 8-HxCDD, 1, 2, 3, 4, 6, 7, 8-HpCDD, other PCDDs, PCDFs and Co-PCBs in paddy soils

HxCDD, 1,2,3,4,6,7,8-HpCDD, other 2,3,7,8-substituted PCDDs, 2,3,7,8-substituted PCDFs, and dioxin-like PCBs in the preserved paddy soils. The highest contributors to TEQ in all samples were 1,2,3,7,8-PeCDD, 1,2,3,6,7,8-HxCDD, and 1,2,3,4,6,7,8-HpCDD. The contribution to TEQ by dioxin-like PCBs was very low in all samples.

After the 1960s, the rate of contribution to TEQ by the following isomers decreased: 1,2,3,6,7,8-HxCDD, 1,2,3,4,6,7,8-HpCDD, and other 2,3,7,8-substituted PCDDs. This decrease can be explained by the reduction in dioxin emissions from PCP. In contrast, the percentage contribution to TEQ by 1,2,3,7,8-PeCDD and 2,3,7,8-substituted PCDFs increased after the 1960s. The former increase can be explained by the use of CNP. However, the increase in the 2,3,7,8-substituted PCDFs cannot be explained in this manner.

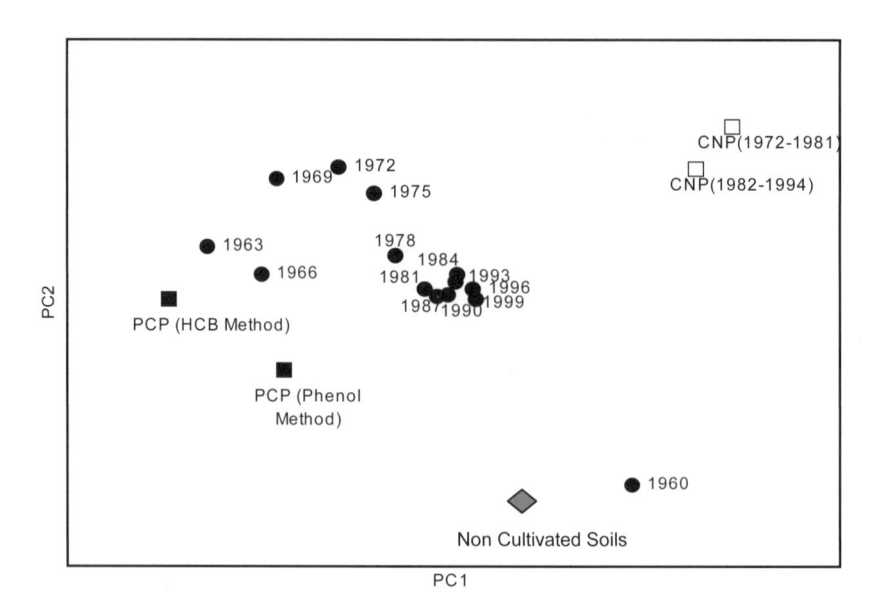

Fig. 9 PCA plot of contribution (%) to TEQ by 2, 3, 7, 8-substituted PCDD/Fs in paddy soils, PCP, CNP and non cultivated soil

In order to evaluate the temporal changes in dioxin sources in paddy soils, we performed a principal-component analysis (PCA; Fig. 9). In our PCA, we used the percentage contributions to TEQ by 2,3,7,8-substituted isomers in PCP synthesized using the HCB and phenol methods, in CNP synthesized before and after 1981, in paddy soils collected from 1960 to 1999, and in currently uncultivated soils (Ministry of the Environment, 1999). We explain the observed changes in PCDD/F sources as follows:
1. Few contributions resulting from impurities in PCP could be found in paddy soil samples collected at 1960. Atmospheric deposition of emissions produced by combustion processes was thought to be the major source of these depositions before PCP was used.
2. The PCDD/F compositions in the paddy soils changed significantly after farmers began using PCP and CNP in the 1960s and the 1970s.
3. After the 1980s, few changes in dioxin compositions could be found because of reduced emissions from PCP and CNP. However, compositions in samples collected from paddy soils over the past 20 years gradually approached the values in uncultivated soils. This suggests that the proportions of PCDD/Fs emitted by combustion processes gradually increased in paddy soils over the past 20 years.

Acknowledgments

The authors thank Drs. Eun, Komamura and Ishizaka of Japan's National Institute for Agro-Environmental Sciences for their dioxin analysis and the use of their soil sample collection.

References

Alcok, R.E., Mclachlan, M.S., Johnston, A.E. and Jones, K.C. (1998) Evidence for the presence of PCDD/Fs in the environment prior to 1900 and further studies on their temporal trends. *Environ. Sci. Technol.* 32: 1580–15877

Cerlesi, S., Domenico, A.D. and Ratti, S. (1989) 2,3,7,8-tetrachlorodibenzo-p-dioxin (TCDD) persistence in the Seveso (Milan, Italy) soil. *Ecotoxicol. Environ. Saf.* 18: 149–164

Hagenmainer, H. and Brunner, H. (1987) Isomers pecific analysis of pentachlorophenol and sodium pentachlorophenate for 2,3,7,8-substituted PCDD and PCDF at sub-ppb levels. *Chemosphere* 16: 1759–1764

Kjeller, L-O., Jones, K.C., Johnston, A.E. and Rappe, C. (1991) Increases in the polychlorinated dibenzo-p-dioxin and -furan content of soils and vegetation since the 1840s. *Environ. Sci. Technol.* 25: 1619–1627

Kjeller, L-O., Jones, K.C., Johnston, A.E. and Rappe, C. (1996) Evidence for decline in atmospheric emissions of PCDD/Fs in the U.K. *Environ. Sci. Technol.* 30: 1398–1403

MacDonald, R.W., Ikonomou, M.G., Paton, D.W. (1998) Historical Inputs of PCDDs, PCDFs and PCBs to a British Columbia Interior lake: the effect of environmental controls on pulp mill emissions. *Environ. Sci. Technol.* 32: 331–337

Maclachlan, M.S., Sewart, A.P., Bacon, J.R. and Jones, K.C. (1996) Persistence of PCDD/Fs in a sludge-amended soil. *Environ. Sci. Technol.* 30: 2567–2571

Masunaga, S., Takasuga, T. and Nakanishi, J. (2001a) Dioxin and dioxin-like PCB impurities in some Japanese agrochemical formulations. *Chemosphere* 44: 873–885

Masunaga, S., Yao, Y., Ogura, I., Nakai, S., Kanai, Y., Yamamuro, M. and Nakanishi, J. (2001b) Identifying sources and mass balance of dioxin pollution in Lake Shinji basin, Japan. *Environ. Sci. Technol.* 35: 1967–1973

Ministry of Environment (1999) Regarding the results of the urgent simultaneous nationwide survey of dioxins. http://www.env.go.jp/en/topic/dioxin/urgent_Nation-wide_survey.pdf, Ministry of the Environment, Japan.

Ministry of Environment and Ministry of Agriculture, Forestry, and Fisheries (2001) Results on the survey of dioxins in cultivated soils and crops, http://www.env.go.jp/water/dojo/no-diox/index.html, Ministry of the Environment, Japan.

Pearson, R.F., Swackhamer, D.L., Eisenreich, S.J. and Long, D.T. (1997) Concentrations, accumulations and inventories of polychlorinated dibenzo-p-dioxins and dibenzofurans in sediments of the Great Lakes. *Environ. Sci. Technol.* 31: 2903–2909

Seike, N., Sakiyama, T., Matsuda, M. and Wakimoto, T. (1994) Bioaccumulations of non-2,3,7,8-substituted PCDFs in fishes of environmental samples. *Organohalogen Compounds* 20: 147–150

Seike, N., Matsumoto, M., Matsuda, M., Kawano, M. and Wakimoto, T. (1998) Distribution and residue patterns of polychlorinated dibenzo-p-dioxins and dibenzofurans in coastal, river and pond water and sediments from Matsuyama, Japan. *Organohalogen Compounds* 39: 97–100

Wakimoto, T., Kannan, N., Ono, M., Tatsukawa, R. and Masuda, Y. (1988) Isomer-specific determination of polychlorinated dibenzofurans in Japanease and American polychlorinated biphenyls. *Chemosphere* 17: 743–750

Yamagishi, T., Miyazaki, T., Akiyama, K., Morita, M., Nakagawa, J., Hori, S. and Kaneko, S. (1981) Polychlorinated dibenzo-p-dioxins and dibenzofurans in commercial diphenyl eater herbicides, and in freshwater fish collected from the application area. *Chemosphere* 10: 1137–1144

Development of a crop-soil database for evaluation of the risk of cadmium contamination in staple crops

Kazuo Sugahara*, Tomoyuki Makino and Yasuhiro Sakurai

National Institute for Agro-Environmental Sciences, 3-1-3 Kannondai, Tsukuba, Ibaraki 305-8604, Japan

Abstract

To prepare for the implementation of the new cadmium standard being developed by the Codex Committee on Food Additives and Contaminants, Japan's Ministry of Agriculture, Forestry and Fisheries has launched several projects to minimize cadmium contamination in crops. The National Institute for Agro-Environmental Sciences has also organized a research consortium to develop a solution for cadmium contamination in staple crops. This paper introduces the framework for an ongoing project to assess the risk of cadmium contamination in staple crops. Soil samples from 767 locations across Japan were collected from April 1, 2000 to March 31, 2003 to analyze cadmium concentrations in crops and soils, and to identify the relevant soil physicochemical properties. A new crop-soil database was then developed to evaluate the risk of cadmium contamination in these crops. Multiple-regression analysis, principal-components analysis, and discriminant analysis all proved to be useful for evaluating the risk of cadmium contamination.

Keywords : cadmium, multivariate analysis, rice, soybean, spinach

1. Introduction

The Codex Alimentarius Commission (CAC) and its Committee on Food Additives and Contaminants (CCFAC) have both begun to develop a new standard for cadmium levels in foods (CAC, 2001, 2002). CAC is the joint food standard committee established by the Food and Agriculture Organization of the United Nations (FAO) and the World Health Organization (WHO), and is the body responsible for compiling standards, codes of practice, guidelines, and recommendations. CCFAC has asked the FAO/WHO Joint Expert Committee on Food Additives (JECFA) to conduct a risk assessment for chronic toxicity of cadmium to humans. JECFA has been mandated to conduct independent risk assessments based on scientific data to support the deliberations by CCFAC.

The activities of CCFAC and JECFA aimed at establishing a new standard for cadmium levels in foods are summarized as follows (Asakura, 2001):

Table 1 Proposed draft maximum levels for cadmium*

Food	ML (mg/kg)	Step	Remarks
Fruit	0.05	3	
Wheat grain	0.2	3	Including bran and germ
Milled rice	0.2	3	
Soybean and peanuts	0.2	3	
Meat of cattle, poultry, pig and sheep	0.05	3	
Molluscs	1.0	3	
Meat of horse	0.2	3	
Vegetables	0.05	3	Excluding leafy vegetables, fresh herbs, stem and root vegetables, fungi, tomatoes and peeled potatoes
Peeled potatoes, stem and root vegetables	0.1	3	Excluding Celeriac
Leafy vegetables, fresh herbs, fungi and celeriac	0.2	3	

* The provisional standard discussed in the 34th CCFAC.

1. In 1988, JECFA conducted a risk assessment on long-term cadmium intake and established a provisional tolerable weekly intake (PTWI) of 7 μg/kg of body weight.
2. In March 1998, CCFAC proposed a provisional standard for the cadmium concentration in foods based on Denmark's proposal.
3. In March 1999, CCFAC promoted the provisional standard proposed by Denmark to the step 3 in the approval process (Table 1) and asked the member countries involved in the process to comment on the standard.
4. In June 2000, JECFA conducted an additional risk assessment for cadmium. They concluded that the existing epidemiological studies on chronic toxicity of cadmium provided insufficient data, and that it was not yet possible to conduct a satisfactory risk assessment of cadmium in humans. JECFA then asked Japan to conduct epidemiological studies on the relationship between the frequency of cadmium-related diseases and the long-term intake of foods that contain cadmium.
5. In March 2001, CCFAC asked JECFA to conduct a risk assessment for cadmium toxicity in preparation for 2003 JECFA meeting, based on the epidemiological studies that Japan had been conducting since 2001. CCFAC agreed that the provisional standard for cadmium would be revised, if necessary, on the basis of JECFA's risk assessment.

To facilitate the risk assessment for cadmium toxicity, the Japanese government agreed to conduct a variety of epidemiological studies targeting the group that was most vulnerable to cadmium toxicity, namely, adult women across Japan (Tsukahara et al., 2003). To prepare for the implementation of the new cadmium standard under discussion by CCFAC, Japan's Ministry of Agriculture, Forestry and Fisheries (MAFF) has launched several projects to minimize cadmium contamination in staple crops that Japan was mandated to investigate by CCFAC, and the National Institute for Agro-Environmental Sciences (NIAES) has organized a research consortium to develop a solution for cadmium contamination in rice, soybean, and spinach.

In this paper, we introduce the framework for an ongoing research project entitled "Technological development for the risk assessment of cadmium contamination in staple crops in terms of cadmium concentration in farm soils", and discuss some of the results obtained thus far.

2. Strategy and an action plan for new cadmium research

In April 2002, NIAES started a new research project in collaboration with the National Agricultural Research Organization, the Akita Agricultural Experiment Station, the Hokkaido Central Agricultural Experiment Station, the Toyama Agriculture Technology Center, the Saitama Agricultural and Forestry Research Center, and the Ibaraki Agricultural Center.

The objectives of this project were (1) to develop methodology for the measurement of bioavailable cadmium in arable soils and (2) to develop a new crop-soil database to permit the evaluation of the risk of cadmium contamination.

The project is currently composed of the following four major topics (Fig. 1):
1. Methodology development for the determination of bioavailable cadmium in arable soils and its *in situ* verification: The relationship between cadmium absorption by rice and

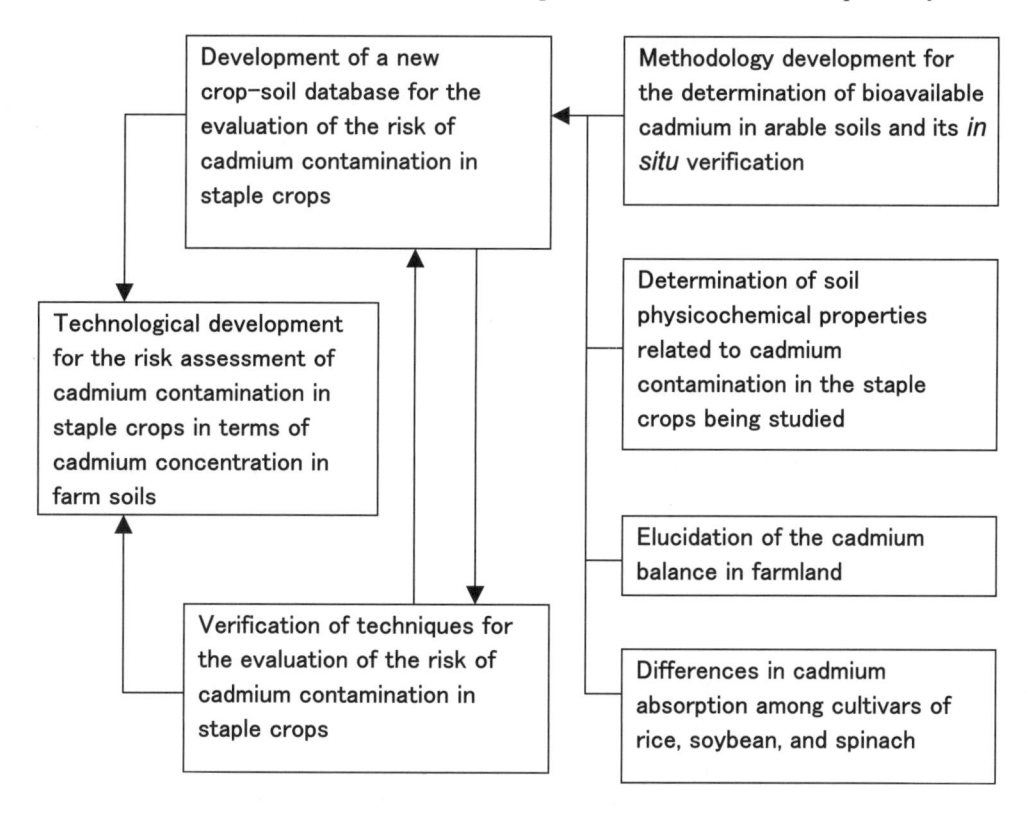

Fig. 1 Strategy and an action plan for new cadmium research

other field crops and the chemical forms of cadmium existing in arable soils is analyzed statistically, and the relevant soil factors that affect the bioavailability of cadmium are determined. After on-site verification testing, a new method will be proposed for the determination of bioavailable cadmium.

2. Determination of soil physicochemical properties related to cadmium contamination in the staple crops being studied: A large number of cadmium concentrations in soils and crops have been collected and accumulated by a variety of projects, such as the National Soil Conservation Survey, and have been integrated into a new crop-soil database to assist in the prediction of the risk of cadmium contamination. Furthermore, data on soil physicochemical properties that are related to cadmium absorption by crops, which were missing from data provided by previous projects, have been determined and added to the new crop-soil database. Using the new database, the relationship between the soil properties that are hypothesized to affect cadmium absorption by crops and the resulting cadmium concentrations in edible plant parts is being statistically analyzed; the soil properties that significantly affect cadmium absorption are being determined.

3. Elucidation of the cadmium balance in farmland: The cadmium loads supplied to farmland by drainage water and by inorganic and organic fertilizers are being analyzed to elucidate the cadmium balance in a model area in Akita Prefecture. In addition, cadmium accumulation is being successively surveyed in long-term field plots that receive both inorganic and organic fertilizers, and the contribution of individual components responsible for the cadmium accumulation is being evaluated.

4. Differences in cadmium absorption among cultivars of rice, soybean, and spinach: Under this topic, provincial agricultural experiment stations and research centers are collaborating with NIAES to identify promising cultivars of the mandated staple crops that have a low ability to absorb cadmium.

By fully exploiting the data in the new crop-soil database, we have been able to establish methodologies to evaluate the risk of cadmium contamination in the aforementioned staple crops. Furthermore, farm lands where the staple crops are cultivated will be classified according to their risk of cadmium contamination in these crops, and a risk map for cadmium contaminations will be created for every crop. On the basis of the risk levels for cadmium contamination, an appropriate and economically feasible technology will be developed to minimize contamination in the crops being studied.

3. Application of multivariate analysis in soybean

The relevant soil physicochemical properties that are closely associated with bioavailable soil cadmium and, hence, the amount of cadmium absorbed by the crops being studied, have been highlighted. First, we developed the aforementioned crop-soil database on cadmium contamination in these crops. The database consists of data on the cadmium concentrations in edible parts of crops, soil cadmium concentrations, and the relevant soil physicochemical properties. The following soil properties were selected as indicators of bioavailable cadmium: 0.1 M HCl-extractable cadmium (soil cadmium), soil pH, pH buffering capacity of the soil,

Table 2 Soil and crop samples collected from 767 locations across Japan

Year	Wheat & barley	Soybean	Spinach	Total
2000	20	66	25	111
2001	14	229	16	259
2002	251	125	21	397
Total	285	420	62	767

total carbon in the soil, phosphate-absorption coefficient, cation-exchange capacity (CEC), exchangeable cadmium, and exchangeable cations in the soils.

Soil and crop samples were collected from 767 locations across Japan during the 2000 to 2002 fiscal years to permit our analysis of cadmium concentrations in crops and soils. The soil properties described in the previous paragraph were also analyzed. The soils, grouped by crop type, are presented in Table 2. Up to the present, we have completed our analysis of cadmium concentrations in the edible part of these crops and the analysis of six soil properties (soil cadmium, soil pH, total carbon, CEC, phosphate-absorption coefficient, and exchangeable cadmium). The data on the samples from 2000 and 2001 have been entered into the crop-soil database. Multivariate analyses (multiple-regression analysis, principal-components analysis, and discriminant analysis) have been applied to the 199 complete data sets for soybean in 2001 using the SPSS statistical-analysis software (SPSS, 1999).

Our results can be summarized as follows:

1. In our multiple-regression analysis, the cadmium concentration in soybean seeds was used as the dependent variable, and six soil properties (soil cadmium, soil pH, total carbon, CEC, phosphate-absorption coefficient, and exchangeable cadmium) were used as independent variables. The analysis revealed a significant correlation ($P<0.001$) between the cadmium concentration in soybean seeds and all six independent variables (coefficient of determination=0.443,

Table 3 Summary of our multiple-regression analysis

R	0.666
R square	0.443
Adjusted R square	0.426
Standard error of the estimates	0.073

Table 4 Coefficients and significance of multiple-regression analysis

	Unstandardized coef. B	Standardized coef. β	Significance
(Constant)	0.314		<0.001
Soil Cd	0.447	0.410	<0.001
Soil pH	-3.85E-02	-0.234	<0.001
Total carbon	-7.24E-04	-0.090	0.272
CEC	-8.08E-04	-0.076	0.442
P-absorption coef.	-3.64E-03	-0.170	0.099
Exchangeable Cd	2.86E-03	0.168	0.021

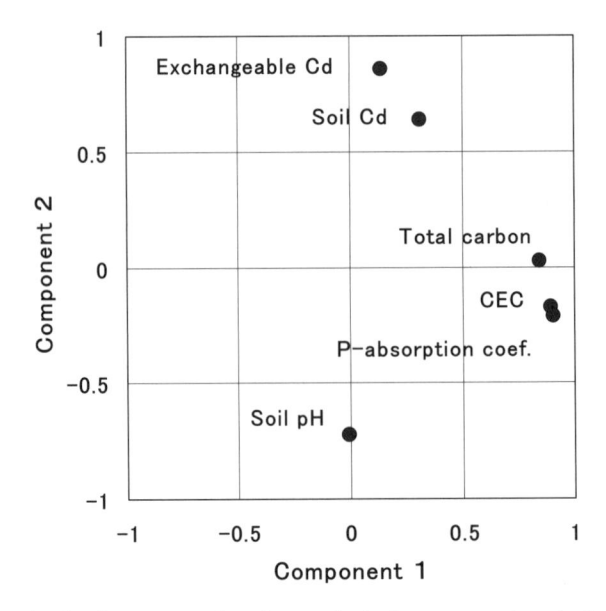

Fig. 2 Component plot of our principal-component analysis

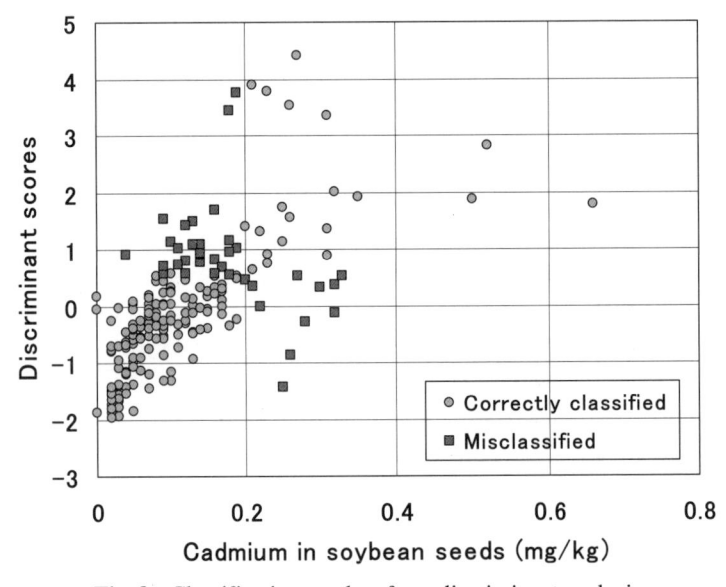

Fig. 3 Classification results of our discriminant analysis

R square in Table 3). In particular, three factors (soil cadmium, soil pH, and phosphate-absorption coefficient) were closely correlated with the cadmium concentration in soybean seeds (Table 4).

2. In our principal-components analysis, soil cadmium, soil pH, total carbon, CEC, phosphate-absorption coefficient, and exchangeable cadmium were selected as variables. The analysis revealed that two components explained about 70% of the total variance: component 1 was related to cadmium sorption sites in the soil, and component 2 was

related to the amount of bioavailable cadmium in the soil (Fig. 2).
3. In our discriminant analysis, the cadmium concentration in soybean seeds was selected as a grouping variable (0.2 ppm was the maximum level in this case), and soil cadmium, soil pH, total carbon, CEC, phosphate-absorption coefficient, and exchangeable cadmium were selected as independent variables. The analysis revealed that about 80% of the original data had been classified properly (Fig. 3). Soil cadmium, soil pH, exchangeable cadmium, and CEC contributed most strongly to the discrimination.

4. Conclusions

Our results can be summarized as follows:
1) Multiple-regression analysis, principal-components analysis, and discriminant analysis were all useful for assessing the risk of cadmium contamination in soybean seeds.
2) Soil cadmium, soil pH, phosphate-absorption coefficient, and CEC were all strongly associated with cadmium concentrations in soybean seeds.
3) About 80% of the original data were classified properly in the discriminant analysis.

Acknowledgments

The authors thank Dr. Hideo Imai and Dr. Tetsuhisa Miwa for their invaluable discussion and advice.

References

Asakura K (2001) Circumstances on the international foods standard for cadmium, *Jpn. J. Soil Sci. Plant Nutr.* 72: 707

CAC (2001) Report of the 33rd session of the Codex committee on food additives as contaminants, Hague, Netherlands

CAC (2002) Report of the 34th session of the Codex committee on food additives as contaminants, Rotterdam, Netherlands

SPSS (1999) SPSS Base 10.0 application guide 189-230, 243-292, 317-358

Tsukahara T, Ezaki T, Moriguchi J, Furuki K, Shimbo S, Matsuda-Inoguchi N, and Ikeda M (2003) Rice as the most influential source of cadmium intake among general Japanese population, *Sci. Total Environ.* 305: 41-51

Estimating methyl bromide emissions due to soil fumigation, and techniques for reducing emissions

Yuso Kobara

National Institute for Agro-Environmental Sciences 3-1-3 Kannondai,
Tsukuba, Ibaraki 305-8604, Japan

Abstract

Methyl bromide (CH_3Br) is an important fumigant used to control soil-borne diseases. About 76 % of industrial CH_3Br is used for this purpose. In the Montreal Protocol, CH_3Br use is to be restricted because it depletes stratospheric ozone, and its use as a soil fumigant is to be phased out entirely by 2005. Currently, only critical and emergency uses are permitted. We studied ways to improve fumigation technologies so as to reduce the required CH_3Br dosage and subsequent emissions while maintaining its effectiveness for disease and weed control. Field measurements showed that the intensity of solar radiation determined changes in emissions, which varied by as much as 1,300% on clear days. Our results suggest possible emissions reductions of nearly 30%; thus, the benefits from the phase-out of CH_3Br may be substantially less than previously estimated. The use of gas-tight films can reduce both the required dosage and subsequent emissions, while increasing retention times in the soil. Shielding the soil from solar radiation can also minimize CH_3Br emission. Emissions could be further reduced by using a photocatalyst: CH_3Br emission was reduced to less than 1% of the amount applied by using sheets containing titanium dioxide photocatalyst.

Keywords : Methyl bromide, soil fumigation, emission reduction, Very Impermeable Films (VIFs), titanium dioxide (TiO_2) photocatalyst

1. Introduction

Methyl bromide (CH_3Br) is an atmospheric trace gas, present in the atmosphere at around 10 parts per trillion (ppt = 10^{-12}) by volume. This gas raises concerns because it is the primary carrier of bromine into the stratosphere, and because the reaction-rate enhancement caused by bromine makes this compound 50 to 60 times more effective than chlorine in destroying stratospheric ozone (Albriton and Watoson, 1992; Kurylo, et al., 1999). Thus, 10 ppt of bromine would be the equivalent of 500 to 600 ppt of chlorine. The present-day amount of organic chlorine in the atmosphere is about 3500 ppt, of which only 550 ppt occurs naturally (Anderson et al., 1989; Montzka et al., 2002). Normally, chemicals such as CH_3Br with a very short lifetime in the atmosphere have a low relative impact on stratospheric ozone because they maintain a low mixing ratio in the atmosphere. Although natural sources dominate the CH_3Br budget, there is a sizeable anthropogenic flux to the atmosphere though its use as a fumigant. Its industrial production is due to be phase out, largely because of its high ozone-depletion potential.

Many natural sources of CH_3Br exist, but there is substantial uncertainty over the quantitative values of their contributions to the total level in the atmosphere. Known sources of CH_3Br include oceanic emissions, biomass burning (Andreae et al., 1996), agricultural fumigations, leaded gasoline, and structural fumigation such as buildings, ships and aircrafts. About 76 % of industrially produced CH_3Br (amounting to 71 Gg yr^{-1} in 1996) is used as an agricultural fumigant against soil-borne diseases and weeds, owing to the chemical's wide spectrum of action. During fumigation, some of the CH_3Br is degraded in the soil, but the remainder is emitted into the atmosphere. Currently, 26.5 Gg yr^{-1} (range: 16 to 48 Gg yr^{-1}) of CH_3Br is believed (Montzka et al., 2002) to be emitted during and after fumigation, and this represents a significant proportion of the total global emissions of 159 Gg yr^{-1} (range: 77 to 293 Gg yr^{-1}; Tables 1, 2, and 3). On the basis of mainly indirect measurements in controlled environments, several authors have reported that the amount of emissions varied with soil type, humidity, pH, organic matter content, and the method of application (Yagi et al., 1993, 1995; Majewski et al., 1995; Yates et al., 1996, 1997; Williams et al., 1997; Wang et al., 1997b, 1998). Direct measurements of emissions from fumigated fields have, however, been rare.

Table 1 Estimating global sales of CH_3Br by use sector (metric ton)

Year	Soil	Post harvest	Structural	Chemical intermediate	Total sales
1984	30,408	9,001	2,166	3,997	45,572
1985	33,976	7,533	2,257	4,507	48,273
1986	36,090	8,332	2,029	4,004	50,455
1987	41,349	8,708	2,923	2,710	55,690
1988	45,131	8,028	3,647	3,804	60,610
1989	47,542	8,919	3,613	2,496	62,570
1990	51,306	8,411	3,234	3,693	66,644
1991	55,079	10,290	1,817	4,071	71,257
1992	57,407	9,855	2,264	2,648	72,174
1993	n/a	n/a	n/a	n/a	72,658
1994	n/a	n/a	n/a	n/a	73,731
1995	n/a	n/a	n/a	n/a	66,778
1996	47,896	13,948	3,993	2,531	68,424

Methyl Bromide Global Coalition 1994, UNEP 1995, and ICF 1996 and 1997
*) Use on perishable not included in 1991 and 1992.
n/a) not available.

Table 2 Histrical consumption[1] of fumigants by end use in Japan (Active ingredient, t)

Year[2]	Methyl bromide	Chloropicrin	D-D	Dazomet	Methyl isothiocyanate	Metam
1981	5069.6	7011.0	6105.0	0.0	149.0	0.0
1982	5555.0	4602.0	6780.0	0.0	63.0	0.0
1983	6427.0	4241.0	7035.0	0.0	110.0	0.0
1984	7374.0	6570.0	10061.0	48.0	254.0	0.0
1985	6852.0	9301.0	6315.0	93.0	268.0	0.0
1986	6893.0	5284.0	5295.0	83.0	207.0	0.0
1987	8220.0	7554.0	5606.0	43.0	79.0	0.0
1988	8337.0	7903.0	8025.0	102.0	173.0	0.0
1989	8502.0	7772.0	9460.0	76.0	277.0	0.0
1990	9881.0	7800.0	7808.0	73.0	223.0	0.0
1991[3]	10418.0	7178.0	8343.0	107.0	257.0	0.0
1992	10312.0	6951.0	7604.0	205.0	214.0	0.0
1993	10623.0	8695.0	6932.0	1064.0	202.0	0.0
1994	11414.0	7656.0	11285.0	1152.0	222.0	10.0
1995	10659.0	8627.0	10311.0	1272.0	209.0	145.0
1996	9330.0	8408.0	8696.0	1615.0	219.0	1183.0
1997	8384.8	9989.0	8122.4	1932.0	222.6	670.2
1998	7755.6	8746.7	13892.1	2620.1	183.2	463.0
1999	6806.6	8275.8	13992.1	2839.7	160.2	311.7
2000	6648.3	9314.2	10699.7	3129.0	221.8	361.8
2001	5049.2	9462.8	8582.5	2316.0	197.8	157.0

Source: Pesticide Handbook, MAFF (Ministry of Agriculture, Forestry and Fisheries).
[1] Consumption = Production+Imports-Exports.
[2] Pesticide Year, for example, 1995 means from October, 1994 to September, 1995.
[3] Standard Year by the Montreal Protocol.

CH_3Br is a major fumigant used in Japan to control soil-borne diseases in crops such as cucumbers, ginger, tomatoes, melons, and green peppers. Its use as a soil fumigant is to be phased out by 2005, but no single chemical or non-chemical alternative has yet emerged to replace it. For now, 1,3-dichloropropene and chloropicrin are generally seen as the best alternatives to CH_3Br for pre-planting fumigation, and their sales are increasing steadily. However, their impacts on the environment and human health are not well understood, so they are considered risky and unacceptable as long-term replacements. It is difficult to adequately satisfy the demand for alternatives to CH_3Br for the preplant fumigation, as only some critical-use exemptions and emergency uses have been permitted.

Restrictions on CH_3Br usage have led to an intensive search for improved technologies to reduce both its dosage and its emission from fumigated plots, while maintaining the chemical's effectiveness for disease and weed control. Improved management practices such as the use of gas-tight films (Wang et al., 1998; Yates et al., 1998), shallow injection in combination with irrigation (Wang et al., 1997a), deep injection (ca. 60-cm depth) (Yates et al.,

Table 3 Emissions of industrially produced CH_3Br and uncertainties in direct CH_3Br emission estimates

Source	Consumption		Emissions		
	Amount (Gg)	Relative to total (%)	*Best estimates (Gg)	Amount (Gg)	Relative to consumption (%)
Industrially produced					
Chemical feedstock	2.4	3	n/a	n/a	n/a
Fumigation - soils	54.8	76	26.5	16 - 48	29 - 88
− durables	10.0	14	6.6	4.8 - 8.4	48 - 84
− perishables	5.5	5	5.7	5.4 - 6.0	98 - 109
− structures	2.2	3	2.0	2.0 - 2.1	91 - 95
Total	71.4	100	46.3	28.2 - 64.5	40 - 90
Gasoline			5	0 - 10	
Oceans			63	23 - 119	
Biomass burning			20	10 - 40	
Wetlands			4.6	2.3 - 9.2	
Salt marshes			14	7 - 29	
Shrublands			1	0.5 - 2	
Rapeseed			6.6	4.8 - 8.4	
Rice fields			1.5	0.5 - 2.5	
Fungus			1.7	0.5 - 5.2	
Peatlands			0.9	0.1 - 3.3	
Total direct			159	77 - 293	

* The estimate for oceanic production is calculated from data in Lobert et al (1995) and Yvon and Butler (1995). Fumigation values are adjusted from UNEP (1994) and biomass burning estimates are taken directly from Mano and Andreae et al (1995). Gasoline emissions are from Penkett et al (1995). Highs and lows for oceanic emissions and total direct emissions represent full possible ranges and probable (rms) ranges.

1997), and the application of organic material or an ammonium thiosulfate fertilizer (Gan et al., 1998a, b), or a soil bacterium such as genus *Rhizobium* (Connell et al., 1997, 1998; Miller et al., 1997), have been shown to reduce CH_3Br emissions in several countries. Mechanical injection methods can reduce the amount of CH_3Br required and its subsequent emissions. However, such techniques are not entirely suitable in Japanese horticultural conditions because field size is generally too small for these methods to apply these methods with machine. In addition, agricultural fields tend to coexist with residential areas where atmospheric pollution arises, and farmers usually apply CH_3Br themselves so they won't have to depend on the use of special applicators. Only soil-surface applications such as the "small-can method", are currently permitted in Japan.

The aims of the work described in the present study are to improve the technologies used in CH_3Br fumigation so as to reduce dosage requirements and emissions from fumigated plots, while maintaining the compound's effectiveness for disease and weed control. These techniques include the use of solar radiation shielding, gas-tight films such as Very Impermeable Films (VIFs), or multilayer sheets containing a TiO_2 photocatalyst.

2. Materials and Methods

Direct measurements of CH_3Br emissions under field conditions were carried out on Hydric Hapludand soils at the National Institute for Agro-Environmental Sciences (NIAES), Tsukuba, Japan. The small-can method was used to apply 32.8 g m^{-2} of CH_3Br to the soil surface (15 m^2: 3 m × 5 m) under conventional polyethylene films (0.05-mm thickness) as well as under the new covering materials described later in this section. Covers were removed 7 d after the treatment. An automated gas chromatography system equipped with flame ionization detectors (GC-FID, Shimadzu GC-14B) and four 7.5-L chambers (diam. 24.5 cm) was used to measure the flux of CH_3Br. The chambers were placed directly on the film or soil surface. In addition, air samples were collected using four STS-25 air samplers (Perkin Elmer) with multi-bed absorbent tubes packed with graphite carbon black (100 m^2 g^{-1}, 60/80 mesh, 190 mg) and a carbon molecular sieve (1200 m^2 g^{-1}, 60/80 mesh, 100 mg). These samples were analyzed by using an automatic thermal desorption system combined with gas chromatography and mass spectrometry (ATD-GC-MS, Perkin Elmer ATD 400 and Varian Saturn 2000 GC/MS). Concentrations of CH_3Br were measured in the air above and below the film and at soil depths of 30, 60, 90, 120, and 150 cm. We used two detectors to measure gas concentrations: the FT-IR-photoacoustic spectrometer (Brül & Kjær, 1301) for CH_3Br, CO_2, and water vapor, and the gas chromatograph for CH_3Br.

The solar radiation shielding with a non-woven, high-density polyethylene fiber sheet (Tyvek®). and the VIFs techniques were evaluated in field experiments from 2 to 12 September 1996 (Trial 1) and from 10 March to 2 April 1997 (Trial 2). We also evaluated reductions in CH_3Br emissions by sheets containing TiO_2 photocatalyst from 5 to 30 May (Trial 3), 26 August to 20 September (Trial 4), and 16 November to 5 December 1998 (Trial 5), and from 4 to 20 August 1999 (Trial 6).

From gas-tight films (VIFs) for CH_3Br, ethylene-vinyl alcohol copolymer (EVOH, C.I. Kasei) was selected based on the criterion that the material also has the maximum transmittance of ultraviolet light (wavelength < 400 nm). TiO_2 photocatalyst (Ishihara Sangyo Kaisha, Ltd., ST-01) was suspended in the acetone solvent, then spread at a rate of ca. 3 g m^{-2} on non-woven, high-density polyethylene fiber sheets (Tyvek®, DuPont), then heat-sealed to the Tyvek® sheet with an EVOH gas-tight film. The sheet therefore consisted of an impermeable layer (top), a photocatalyst layer, and a support layer (bottom). In laboratory experiments, this multilayer sheet was set up in the center of the separable chamber (10 cm in effective irradiation diameter, with upper and lower chamber volumes of ca. 400 and 280 mL, respectively), then distilled water (1 mL) and CH_3Br gas (2.5 mL at room temperature and atmospheric pressure) were introduced into the lower chamber. Irradiation was provided by a 500-W Xe arc lamp (Wacom, WXS-105C-5) that approximated to the AM (Air Mass) 1.5 G (where G stands for "global" and includes both direct and diffuse radiation) standard solar radiation at the room temperature (Kobara et al., 2002).

3. Results and Discussion

3.1. Solar radiation shielding

In a previous study (Kobara et al., 1997), direct measurements under field conditions showed that the rate of CH_3Br emission depended strongly on the level of solar radiation, on temperature, and on the CH_3Br concentration below the film. The results indicated that fumigation on cloudy days or around sunset is a simple but effective method for reducing CH_3Br emissions into the atmosphere. Providing additional shielding against solar radiation can be even more effective. To reduce emissions into the atmosphere by restraining temperature increases during application, we improved the application method by using conven-

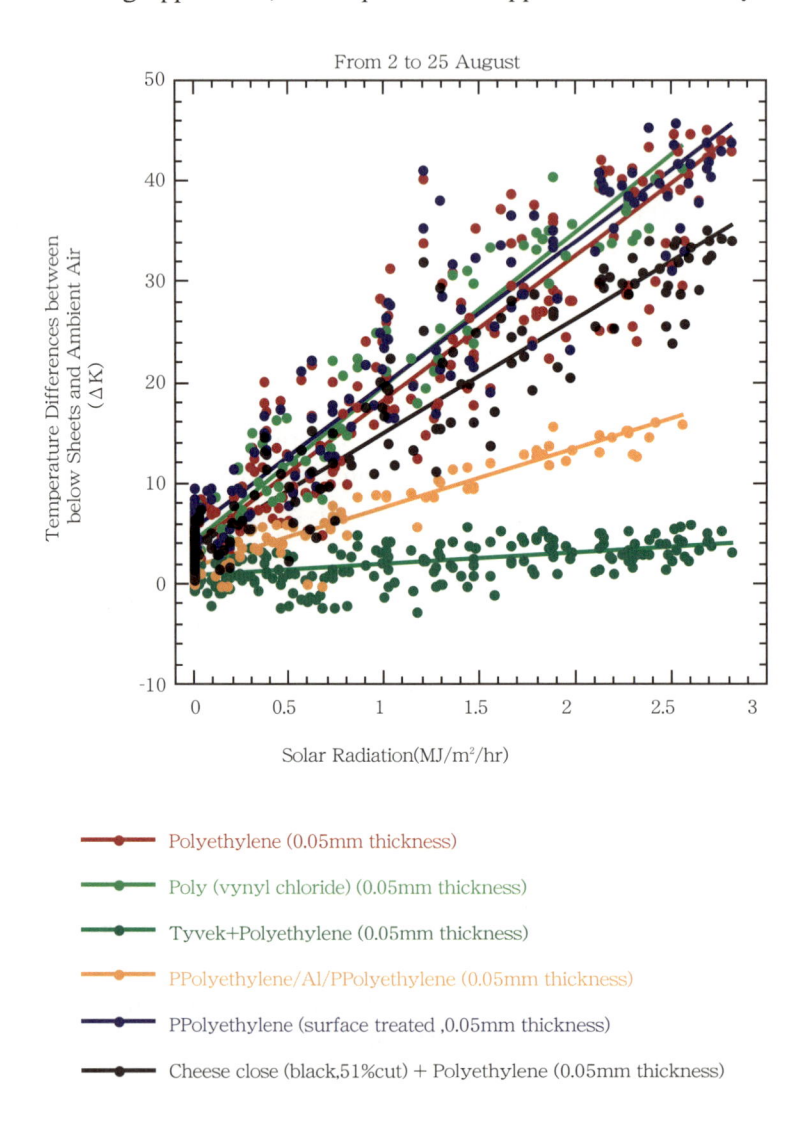

- Polyethylene (0.05mm thickness)
- Poly (vynyl chloride) (0.05mm thickness)
- Tyvek+Polyethylene (0.05mm thickness)
- PPolyethylene/Al/PPolyethylene (0.05mm thickness)
- PPolyethylene (surface treated ,0.05mm thickness)
- Cheese close (black,51%cut) + Polyethylene (0.05mm thickness)

Fig. 1 Evaluation of typical covering sheets regarding green house effect

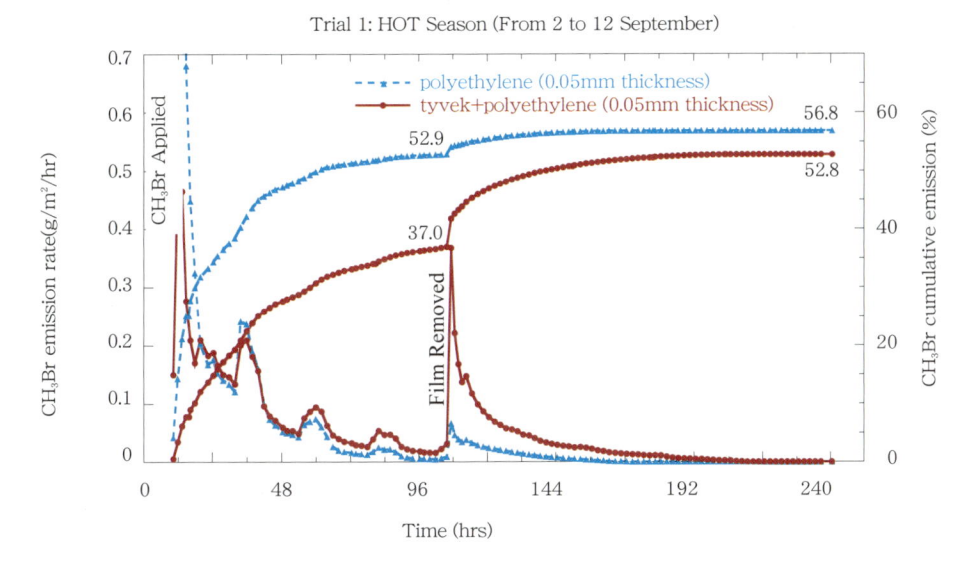

Fig. 2 CH_3Br emission rates $(g/m^2/hr)$ and cumulative emission to the atmosphere as fraction of the applied amounts. Data points are means of each two measurements. Soil was covered with a film for 7 d and then removed

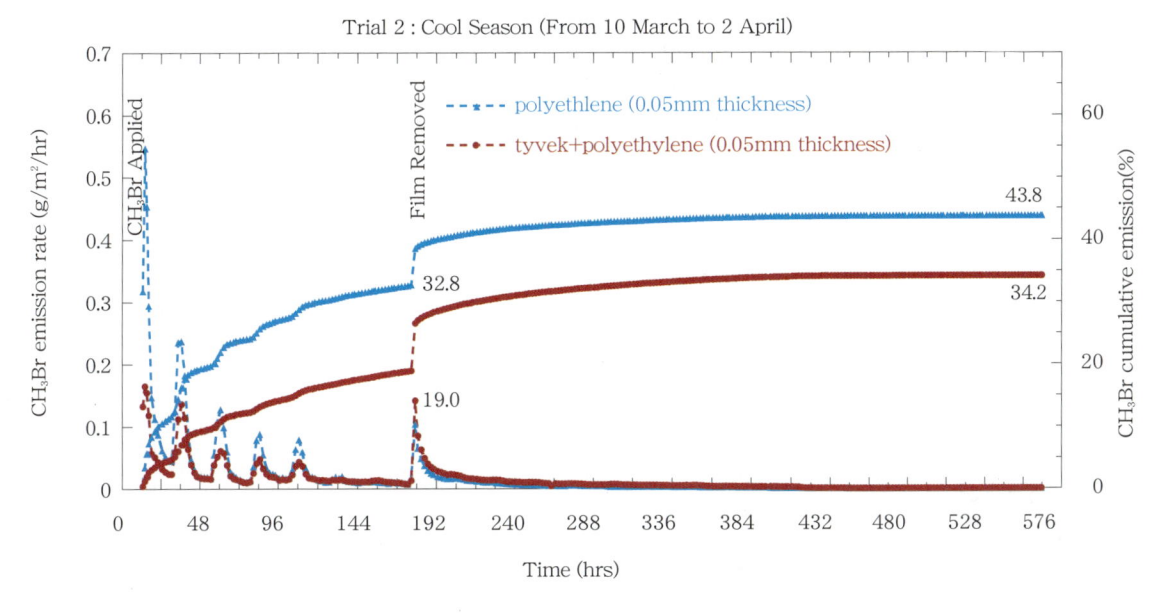

Fig. 3 CH_3Br emission rates $(g/m^2/hr)$ and cumulative emission to the atmosphere as fraction of the applied amounts. Data points are means of each two measurements. Soil was covered with a film for 7 d and then removed

tional polyethylene (PE) and polyvinyl chloride (PVC) films combined with a non-woven, high-density polyethylene fiber sheet (Tyvek®). When used as a cover sheet, Tyvek® is able to shield the soil surface, and PE and PVC films from solar radiation by diffuse reflection (Kobara et al., 1997).

Temperatures below the films other than the Tyvek® sheet varied widely during the day, with changes in the intensity of solar radiation. In contrast, temperatures were nearly equal to ambient temperatures when Tyvek® sheeting was used (Fig. 1). The results suggested that using Tyvek® sheeting reduced emissions by 30 % to 42 % during fumigation, and by 7 % to 22 % on the whole as compared with the control (without Tyvek®). This technique was more effective in cool weather (Trial 2) than in hot weather (Trial 1) (Fig. 2, 3), further, had higher CH_3Br concentration below the film during fumigation. The Tyvek® sheet can easily be obtained and can be re-used repeatedly. The problem of waste disposal is also smaller because VIFs that contain chlorine, such as polyvinylidene chloride (PVDC), are especially difficult to destroy by burning, whereas Tyvek® is easy to burn and generates no chlorine emissions. Moreover, solar radiation shielding technique has the advantage of being cheaper than VIFs technique.

3.2. VIFs (Very Impermeable Films)

We found that using a gas-tight film (Orgalloy®, 0.05-mm thickness Elf Atochem) during surface applications considerably reduced emissions (to a net loss of 7.6% of the amount applied) during the 7 days after an application. However, emissions increased greatly soon after we removed the film, and eventually totaled 33% of the amount applied by the end of the study period (Fig. 4). The total emission is thus similar to that with conventional films such as polyethylene. The standard dose of CH_3Br used in Japan varies from 15 to 30 g m^{-2}, which is near the minimum effective level, and it is difficult to dramatically reduce dosages

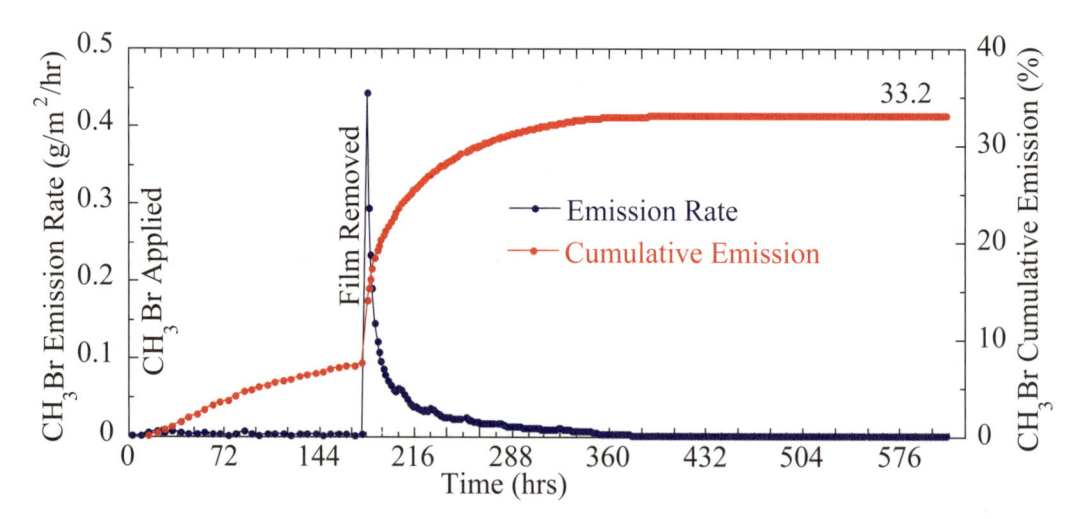

Fig. 4 Reduction techniques of CH_3Br from soil fumigation with a gas-barrier film (Orgalloy film, 0.05 mm thickness, elf atochem), From 11 April to 6 May

using very impermeable films alone. In other countries (Bell et al., 1996), the recommended dosage rates for commercial applications range from 50 to 100 g m^{-2}.

3.3. Multilayer sheet containing TiO_2 photocatalyst

One purpose of our study was to develop and evaluate a new multilayer sheet for use in surface application of CH_3Br. We assumed that emissions would be significantly reduced if CH_3Br degradation was enhanced by the presence of a photocatalyst, although this approach has not been well documented at this time.

We chose TiO_2 as the photocatalyst for several reasons. First, TiO_2 is a photo-semiconductor that causes a redox reaction at its surface in response to ultraviolet (UV) radiation (less than 400 nm). In this reaction, the compound generates active forms of oxygen, such as superoxide anions (O_2^-) and hydroxide (OH) radicals. These compounds degrade CH_3Br into carbon dioxide (CO_2), hydrogen bromide (HBr), and water. In addition to having a strong photocatalytic action, TiO_2 is considered to be environmentally safe.

Both EVOH (0.060 mm: low-density polyethylene / ethylene-vinyl alcohol co-polymer / low-density polyethylene) and fluorinated polymer films (0.050 mm) have excellent barrier properties and UV-transmittance, but we chose EVOH because of the ease of heat-sealing this material. Tyvek® sheet was chosen as the support layer for the TiO_2. Tyvek® reflects nearly 100% of diffuse reflection by means of its ultra-fine polyethylene fibers, and offers good gas permeability.

Concentrations of CH_3Br at the beginning of the test were about 6000 ppm in the lower chamber, but decreased to near-zero levels within 48 h after irradiation. The degradation products were identified as CO_2 and HBr. As the generated HBr was neutralized immediately

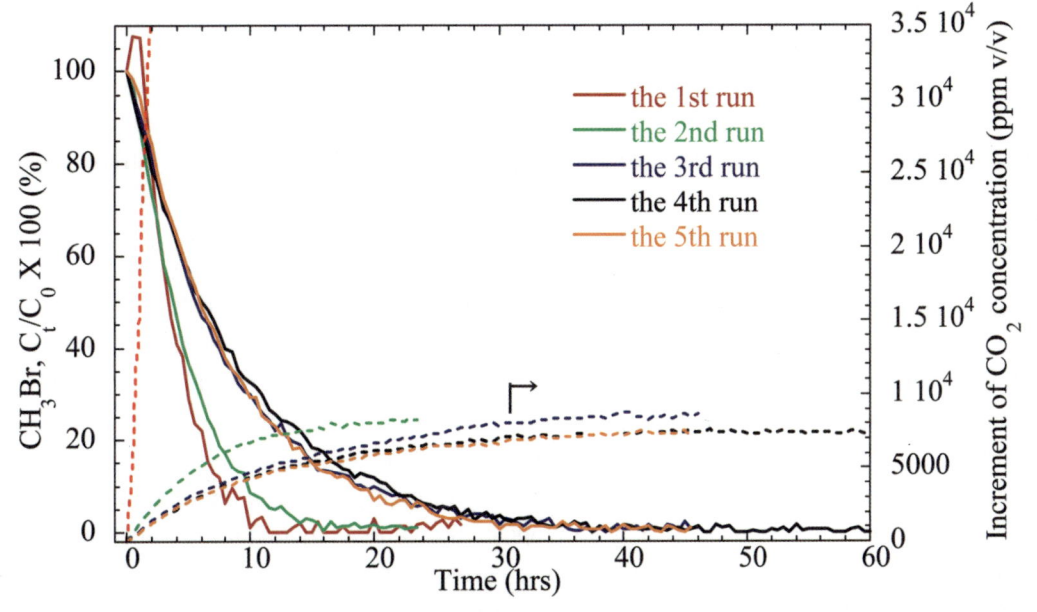

Fig. 5 Performance of the multilayer sheet containing a photocatalytic layer (TiO_2) by repeated runs. CH_3Br degradation into CO_2 by photooxidation (TiO_2 with 10 % PTFE particles as a binder) C_0 = ca. 6300 ppm v/v

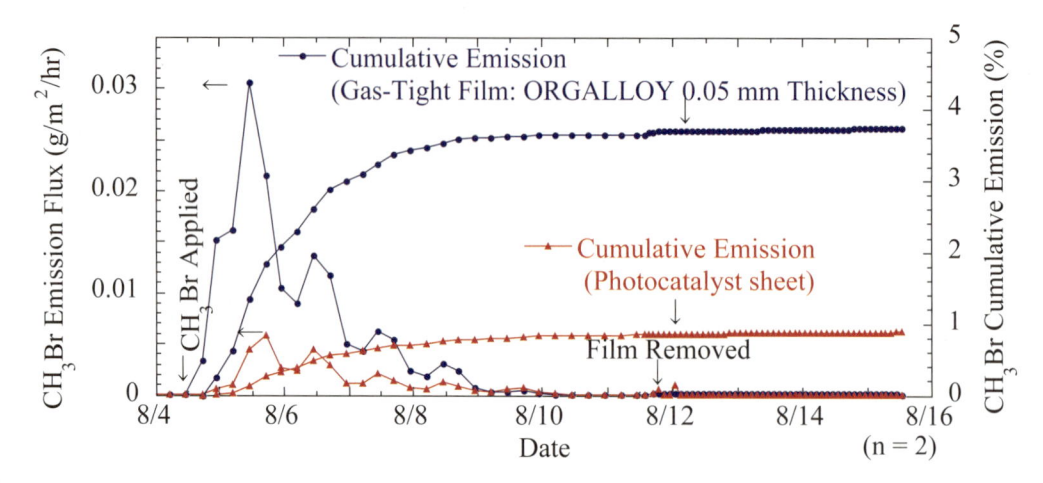

Fig. 6 Comparison of CH_3Br emission from soil fumigation, the multilayer sheet containing a photocatalytic layer (TiO_2) and a gas-tight film. Data points are means of each two measurements. Soil was covered with a film for 7 d and then removed

by the soil under field conditions, most of the remaining CH_3Br recovered in the field at the end of the experiment was found near the soil surface and the multilayer sheet.

The ability of the multilayer sheet to decompose CH_3Br decreased with repeated use (Fig. 5; up to five uses) because of detachment of TiO_2 from the sheet. It was possible to prevent this detachment by mixing the TiO_2 with ca. 10% fine poly(tetrafluoroethylene) particles as a binder (Kobara et al., 1998, 1999a, 1999b, 1999c).

Although decomposition and removal rates of CH_3Br are slow and depend on levels of solar radiation, CH_3Br concentrations below the sheet declined rapidly while the sheet covered the field (7 or 9 days). Just before the removal of the sheet, CH_3Br concentrations between the sheet and the soil surface decreased to a few ppm (v/v) with the multilayer sheet, but remained high (1000 ppm [v/v]) with the Orgalloy® gas-barrier film (Kobara et al., 1999, 2001).

Our experiments also showed that CH_3Br emission decreased to less than 1% of the applied amount in the treatment that used the multilayer sheet containing TiO_2, but decreased to only about 57% and 33% of the applied amount with a polyethylene sheet (0.05 mm thickness in the traditional method) and a gas-barrier film, respectively (Fig. 6). Moreover, CH_3Br concentrations below the multilayer sheet and the gas-tight film were similar until the middle of the fumigation period. This indicates that under field conditions, the use of a multilayer sheet may not greatly reduce the efficacy of methyl bromide fumigation. The multi-layer sheet can also be re-used easily and repeatedly without any major modifications in the current approach to soil-surface application. Furthermore, there are minimal problems in disposing of the sheet. We therefore believe that the technique can substantially reduce CH_3Br emissions and that multilayer sheets containing TiO_2 hold promise for commercial use. Despite the promise of this approach, we must still find ways to improve application methods and must still seek environmentally acceptable and agriculturally effective chemical alternatives to CH_3Br.

Acknowledgments

The studies reported here form part of a research project sponsored by Japan's Ministry of the Environment.

References

Albritton, D. L. and Watson, R. T. (1992) Methyl bromide and the ozone layer: A summary of current understanding, Methyl Bromide: Its Atmospheric Science, Technology and Economics, Montreal Protocol Assessment Supplement, edited by Watson, R. T., Albritton, D. L., Anderson, S. O. and Lee-Bapty, S. United Nations Environmental Programme (UNEP), United Nations Headquarters, Ozone Secretariat: P.O. Box 30552, Nairobi, Kenya

Anderson, J. G., Bruhne, W. H., Lloyd, S. A., Toohey, D. W., Sander, S. P., Starr, W. L., Loewenstein, M. and Podolske, J. R. (1989) Kinetics of O_3 destruction by ClO and BrO within the Antarctic vortex: an analysis based on in situ ER-2 data. *J. Geophys. Res.*, 94: 11480–11520

Andreae, M. O., Atlas, E., Harris G. W., Helas, G., de Kock, A., Koppmann, R., Mano, S., Pollock, W. H., Rudolph, J., Scharffe, D., Schebeske, G. and Welling, M. (1996) Methyl halide emissions from savanna fires in southern Africa. *J. Geophys. Res.*, 101: 23603–23613

Bell, C. H., Price, N. and Chakrabarti, B. (1996) The Methyl Bromide Issue; John Wiley &Sons

Connell Hancock, T. L., Joye, S. B., Miller, L. G. and Oremland, R. S. (1997) Bacterial Oxidation of Methyl Bromide in Mono Lake, California. *Env. Sci. Technol.*, 31: 1489–1495

Connell Hancock, T. L., Costello, A. M., Lidstrom, M. E. and Oremland, R. S. (1998), Strain IMB-1, A Novel Bacterium for the Removal of Methyl Bromide in Fumigated Agricultural Soils. *Appl. Environ. Microbiol.*, 64: 2899–2905

Gan, J., Yates, S. R., Papiernik, S. K. and Crowley, D. (1998a) Application of organic amendments to reduce volatile pesticide emissions from soil. *Environ. Sci. Technol.*, 32: 3094–3098

Gan, J., Yates, S. R., Becker, J. O. and Wang D. (1998b) Surface amendment of fertilizer ammonium thiosulfate to reduce methyl bromide emission from soil, *Env. Sci. Technol.*, 32: 2438–2441

Kobara, Y., Inao, K. and Ishii, Y (1997) Reducing Emission of Methyl Bromide from Soil Fumigation: Effect of Shielding Solar Radiation with Non-Woven High-Density Polyethylene Fiber Sheet. 1997 Annual International Research Conference on Methyl Bromide Alternatives and Emissions Reductions, 101.1–101.2. *San Diego. USA.*

Kobara, Y., Inao, K. and Ishii, Y (1998) Reducing emission of methyl bromide from soil fumigation: effect of a sheet containing titanium dioxide. 1998 Annual International Research Conference of Methyl Bromide Alternatives and Emissions Reductions, 100.1–100.2 Orlando. USA

Kobara, Y. (1999a) Use of Photocatalysis for Agriculture. *Farming Japan* 33: 41–44

Kobara, Y., Ishii, Y., Ishihara, S. and Inao, K. (1999b) Reducing Methyl Bromide Emissions from Field with a Sheet Containing Titanium Dioxide. 1999 Annual International Research Conference on Methyl Bromide Alternatives and Emissions Reductions, 88.1-88.2 San Diego. USA.

Kobara, Y. (1999c) Reducing methyl bromide emission with a sheet containing titanium dioxide. Methyl Bromide Alternatives Newsletter, United State Department of Agriculture Jun.

Kobara, Y., Ishii, Y., Eun, H., Ishihara, S. and Inao, K. (2001) Estimation of some techniques of reducing methyl bromide emission from soil fumigation under the Japanese horticultural conditions. *Organohalogen Compounds* 54: 238-240

Kobara, Y., Ishihara, S., Ohtsu, K., Horio, K. and Endo, S. (2002) Preliminary studies on the photocatalysis of polychlorinated dibenzo-p-dioxins on soils surface. *Organohalogen Compounds* 56: 429-432

Kurylo, M. J. et al. in Scientific Assessment of Ozone Depletion: 1998 (ed. Ennis, C. A.) (World Met. Org., Geneva, 1999)

Lobert, J. M., Butler, J. H., Montzka, S. A., Geller, L. S., Myers, R. C. and Elkins, J. W. (1995) A net sink for atmospheric CH_3Br in the East Pacific Ocean. *Science.* 267: 1002-1005

Majewski, M. S., McChesney, M. M., Woodrow, J. E., Pruger, J. H. and Seiber, J. N. (1995) Aerodynamic measurements of methyl bromide volatilization from tarped and nontarped fields. *J. Environ. Qual.*, 24: 742-752

Miller, L. G., Connell, T. L., Guidetti, J. R. and Oremland, R. S. (1997) Bacterial Oxidation of Methyl Bromide in Fumigated Agricultural Soils. *Appl. Environ. Microbiol.*, 61: 4346-4354

Manö, S. and Andreae, M. O. (1994) Emission of methyl bromide from biomass burning. *Science.* 263: 1255-1258

Montzka S. A. et al., in Scientific Assessment of Ozone Depletion: 2002 (ed. Ennis, C. A.) (World Met. Org., Geneva, 2003)

Penkett, S. A., et al. (1995) Chapter 10-Methyl Bromide, in Scientific Assessment of Ozone Depletion: 1994, edited by Ennis, C. A. (WMO: Geneva)

UNEP. (1994) 1994 Report of the Methyl Bromide Technical Options Committee, 1995 Assessment, UNEP Montreal Protocol on Substances that Deplete the Ozone Layer, Kenya, UNEP.

Wang, D., Yates, S.R., Ernst, F. F., Gan, J., Gao, F. and Becker, O. J. (1997a) Methyl bromide emission reduction with field management practices. *Environ. Sci. Technol.*, 31:3017-3022

Wang, D., Yates, S. R., Ernst, F. F., Gan, J. and Jury, W. A. (1997b) Reducing methyl bromide emission with high-barrier film and reduced dosage. *Environ. Sci. Technol.*, 31: 3686-3691

Wang, D., Yates, S. R., Gan, J. and Jury, W. A. (1998) Temperature effect on CH_3Br volatilization: Permeability of plastic cover films. *J. Environ. Qual.*, 26: 821-827

Williams, J., Yang, N. and Cicerone R. J. (1997) Summary of measured emissions of methyl bromide from agricultural field fumigations from six sites in Irvine, California. Presented at the 1997 Methyl Bromide State of the Science Workshop - Summary, Methyl Bromide Global Coalition, DoubleTree Hotel, Monterey, CA, June 10-12, 1997, Abstract, pg. A3-30

Yagi, K., Williams, J., Wang, N. Y. and Cicerone, R. J. (1993) Agricultural soil fumigation as a source of atmospheric methyl bromide. *Proc. Natl. Acad. Sci.*, 90: 8420

Yagi, K., Williams, J., Wang, N. Y. and Cicerone, R. J. (1995) Atmospheric methyl bromide (CH_3Br) from agricultural soil fumigations. *Science,* 267: 1979-1981

Yates, S. R., Gan, J. and Ernst, F. F. (1996) Methyl bromide emissions from a covered field. III. Correcting chamber flux for temperature. *J. Environ. Qual.*, 25: 892-898

Yates, S.R., Gan, J., Wang, D. and Ernst, F. F. (1997) Methyl bromide emissions from agricultural fields: Bare-soil deep injection. *Environ. Sci. Technol.*, 31: 1136–1143

Yates, S.R., Wang, D., Gan, J., Ernst, F. F. and Jury, W. A. (1998) Minimizing methyl bromide emissions from soil fumigation. *Geophys. Res. Lett.*, 25: 1633–1636

Yvon, S. A, and Butler, J. H. (1996) Uncertainties in the effect of the ocean on the atmospheric lifetime of CH_3Br. *Geophys. Res. Lett.*, 23: 53–56

Development of technologies for assessing nutrient losses in agricultural ecosystems in Korea

Pil-Kyun Jung, Yong-Seon Zhang, Seung-Ho Hur, Kee-Kyung Kang and Myung Chul Seo

National Institute of Agricultural Science and Technology, Suwon 441-707, Korea

Abstract

Nutrient losses, especially nitrogen and phosphorus, in agricultural runoff can contaminate surface and ground water, leading to eutrophication. Thus, erosion control is crucial to minimizing nutrient losses from agricultural land. Assessments of various erosion-control practices were carried out under various cropping systems, soil management practices, and slope conditions by means of a lysimeter study and under artificial rainfall. Soil and nutrient losses were monitored in a small agricultural field to evaluate the soil conservation practices. Nutrient losses occur in runoff and leachate (dissolved nutrients) and in sediments (particulate nutrients). Dissolved nitrates accounted for the majority (about 90%) of nitrate transport within the soil. Particulate phosphate in sediments represented the majority (60% to 67%) of phosphate transport. Recently, engineering and agronomic erosion-control practices have been used to reduce erosion problems in fields on slopes. These practices reduced soil loss, runoff, and nutrient loss to 1/6, 1/2, and 1/3 their original levels, respectively. Bioavailable particulate phosphate in sediments represents a variable but long-term source of phosphate for algae. Dissolved nitrate and phosphate are immediately available for algal uptake, so reducing fluxes of these nutrients should also reduce the risk of eutrophication.

Keywords : environment assessment, nutrient loss, soil erosion, eutrophication

1. Introduction

Nutrient losses in runoff from agricultural land can accelerate the eutrophication of surface water. Agricultural activities, particularly the use of agrochemicals, have been recognized as important non-point sources of surface and subsurface water contamination (Parry, 1998; Lal et al., 1998). Nutrients carried by eroding sediments and by runoff may degrade surface-water quality, whereas those leached into the soil and through the crop's root zone by infiltration may eventually contaminate ground water (Ng Kee Kwong et al., 2002; Uusitalo et al., 2000).

Korea is a mountainous country, and two-thirds of its total land occupies steep mountainous terrain. In addition, land on slopes is very vulnerable to soil erosion owing to the soil's coarse texture and the concentrated heavy rainfall that occurs during the summer.

Chemical fertilizers, pesticides, and livestock waste are the main factors responsible for pollution of streams and lakes, which is often followed by eutrophication. High nitrogen and phosphorus surpluses due to increasing livestock production and increasing use of inorganic fertilizers have had an especially detrimental effect on water quality in Korea. This paper describes Korean agricultural practices in view of the nutrient losses from and soil conservation practices being used in farmland on slopes. Recent research relating to nutrient loss and balances is briefly presented.

2. Livestock Wastes and Soil

Based on data for 2001, the overall production of livestock wastes in Korea was approximately 32 million tons (dry-matter basis). Most of this represented cattle and pig wastes, which were the main sources of soil and water pollution in Korea. The application of phosphorus from livestock sources (Table 1) to agricultural soils leads to high loads of nitrogen and phosphorusin in the soil.

In a benchmark monitoring study, paddy and upland soils were surveyed to detect long-term changes in available soil phosphate (Table 2). The average levels of available phosphate increased gradually in paddy fields, but increased rapidly in upland fields. In the upland fields, some farmers applied more livestock waste than inorganic fertilizer. There are many practical reasons for using livestock wastes in place of inorganic fertilizers, particularly

Table 1 Nirogen and phosphate in the live-stock wastes (20001)

	Live stock	Nitrogen	Phosphate
	Thousand ton		
Cattle	14600 (46)	68	62
Pig	13400 (42)	88	80
Poultry	3900 (12)	40	38
	31900 (100)	214	180

Table 2 Soil available phosphate with year in Korea

	'50	'60	'70	'80	'90	'00
	mg / kg					
Paddy	69	60	88	107	128	136
Upland	152	114	201	231	538	547

Table 3 Distribution of agricultural land uses with slope in Korea

Slope (%)	Paddy field		Upland and Orchard	
	Area (10³ ha)	Rate (%)	Area (10³ ha)	Rate (%)
0–2	550	42.7	96	9.6
2–7	443	34.1	288	28.9
7–15	215	16.7	381	38.2
15–30	45	3.5	201	20.2
>30	35	2.7	71	7.1
Total	1,288	100.0	997	100.0

because of their lower cost and greater availability. However, in some countries such as Korea, where there has been rapid growth in livestock industries, the increased use of livestock waste has caused severe environmental problems (Han and Jung, 1995).

Korea has a total area of 9.930 million ha, of which 23% is classified as agricultural land, 65% as forest, and 12% as "other" uses. Most paddy fields are concentrated in lowlands with a slope of less than 7% and an elevation of less than 100 m above sea level (Table 3). Thus, paddy fields are generally found on flatter land with gentler slopes than are encountered in upland fields. The management of sloping upland fields requires special care to prevent soil erosion (Jung and Oh, 1995).

3. Soil and Water Conservation

We conducted a lysimeter study to determine the effect of soil loss, runoff, and leachate in an upland red pepper field from 1998 to 2001. The soil textures around the lysimeters were sandy loam, loam, clay loam, and clay. Slopes averaged 15%, with a 5-m slope length and a 1.2-m lysimeter depth. The annual soil loss for the various soil textures increased with increasing clay content (Table 4), with one exception: the annual soil loss was lowest in the clay plot, with a 41.1% clay content, owing to the cohesion of the soil.

The effects of conservation practices on soil loss and runoff are shown in Table 5. Conservation practices used in this study included waterways, bench terraces, grass strips, and stone barriers. After these practices were implemented, soil loss and runoff decreased drastically compared with the levels that resulted from the use of conventional practices. We observed rills and gully erosion after heavy rainfall in this area, but conservation practices reduced soil loss to about 1/6 the levels observed from 1995 to 1999, and runoff decreased to

Table 4 Soil losses with different soil texture in the lysimeters (redpepper, 1998~2001)

	Sandy loam	Loam	Clay loam	Clayey
Clay contents (%)	6.5	16.4	28.0	41.1
Soil Loss (kg/10m²/yr)	23.0	31.0	36.0	20.2
Runoff ratio (%)	11.7	15.2	17.4	18.0
Leaching ratio (%)	28.4	25.2	23.3	21.5

Table 5 Soil loss and runoff at the alpine vegetable areas as affected by conservation practices during the research period from 1995 to 1999

Item	Conventional Practices	Conservation Practices
Soil loss (ton/ha)	135	25
Runoff (ton/ha)	1,614	892
Runoff ratio (%)	56.4	30.2

Slope steepness; 25%, Slope length; 200m, Sea level; 800m, Soil texture; silt loam

1/2 these levels (Jung and Oh, 1995).

4. Nutrient Losses in Uplands

We also conducted a study to assess nutrient losses in runoff, leachate, and sediment from a sandy loam soil on a 15% slope. Results were obtained by using lysimeters, and were averaged over 5 years (from 1997 to 2001). Runoff and leachate represented dissolved nutrients, whereas sediments represented clay particles bearing nutrients in particulate form. Dissolved nitrate in the runoff and leachate (Table 6) accounted for the majority (about 90%) of nitrate transport within the soil profile. Particulate phosphate in sediments accounted for the majority (60% to 67%) of phosphate transport (Jung and Oh, 1995; Sharpley et al., 1993).

In the lysimeters, nitrate concentrations in the runoff and leachate decreased during the growing season (Fig. 1). Red pepper was transplanted on 5 May 1997. Nitrogen fertilizer was applied at a rate of 11 kg/10a for a basal dressing in mid-May and at 17 kg/10a for a topdressing in mid-August. Nitrate concentrations were very high after the basal dressing,

Table 6 Nutrient loss in runoff, leachate and sediment from sandy loam with 15% slope in the lysimeter, averaged over 5 years from 1997 to 2001

	Chinese Cabbage	Soybean	Red pepper
	No_3-N (g/10m^2)		
Runoff	299	110	165
Leachate	317	110	292
Sediment	2	2	2
Sum	618	222	459
	P_2O_5 (g/10m^2)		
Runoff	25	10	21
Leachate	12	7	8
Sediment	62	36	44
Sum	99	53	73

Runoff　　$No_3-N : P_2O_5 : K = 1-55 : 0-1.2 : 1-38$ mg/kg
Leachate　$No_3-N : P_2O_5 : K = 1-34 : 0-0.5 : 2-12$ mg/kg
Sediment　$No_3-N : P_2O_5 : K = 1-5 : 75-212 : 50-141$ mg/kg
Sum　　　$No_3-N : P_2O_5 : K = 20-27 : 90-142 : 124-195$ mg/kg

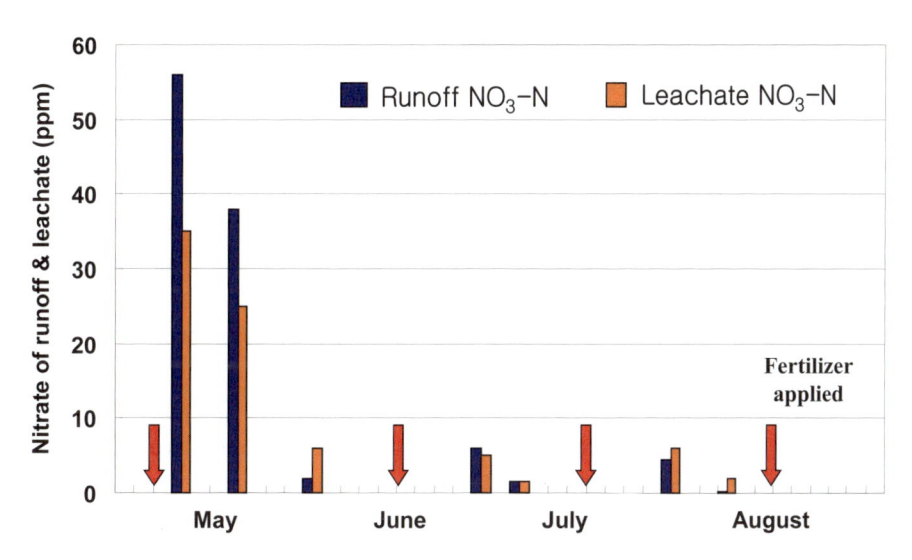

Fig. 1 Nitrate in runoff and leachate during the growing season in lysimeter

but had decreased drastically by mid- to late May. After their initially high levels, nitrate concentrations in runoff and leachate remained below 7 ppm throughout the growing season. Dissolved nitrates transported in runoff and leachate had the strongest impact on surface and groundwater pollution early in the season, after transplanting and application of the basal dressing.

In Korea, vegetables grown in mountainous areas are cultivated as monocultures from the middle of May until early September. The main vegetables cultivated in these fields are Chinese cabbage, radishes, and cabbage, all of which are suitable crops for growing at cool temperatures and high altitudes during the summer. Soil conservation practices were used to control the serious erosion problems that arise in these fields in the northeastern part of South Korea. These practices reduced nutrient losses to about 1/3 of the levels observed from 1995 to 1999 (Jung and Oh, 1995).

5. Conclusions

Our studies have shown that nitrogen and phosphorus losses due to runoff and leaching in farmland on steep slopes represent the main sources of nutrient contamination of surface water and groundwater. Bioavailable particulate phosphates in sediment represent a variable but long-term source of phosphate for algae, and both dissolved nitrate and particulate phosphate are immediately available for algal uptake in agricultural ecosystems, leading to an increased risk of eutrophication of surface water. Appropriate land use, fertilizer management, and soil conservation practices, combined with environmentally sound soil- and water-management practices, can help to reduce nutrient losses and soil erosion, can increase soil productivity, and can protect the quality of surface water and groundwater.

References

Han, J. D. and Jung, K. W. (1995) Managing agricultural wastes. In Colloquium on Sustainability of Agriculture and Conservation of Environment. Association of Korean Agricultural Science Societies. Suwon, Korea: 55–48

Jung, P. K. and Oh, S. J. (1995) Soil and water conservation of sloped farmland in Korea. In Proceedings of Soil Conservation and Management for Sustainable Slope Land Farming. Ping-tung, Taiwan: 15–2–15–15

Lal, R. (1998) Soil erosion impact on agronomic productivity and environmental quality. *Crit. Rev. Plant Sci.* 17: 319–464

Kwong, K. F., Bholah, A., Volcy, L. and Pynee, K. (2002) Nitrogen and phosphorus transport by surface runoff from a silty clay loam soil under sugarcane in the humid tropical environment of Mouritius. *Agric. Ecosyst. Environ.* 91: 147–157

Parry, R. (1998) Agricultural phosphorus and water quality. A US Environmental Protection Agency Perspective. *J. Environ. Qual.* 27: 258–260

Sharpley, A. N., Daniel, T. C. and Edwards, D. R. (1993) Phosphorus movement in the landscape. *J. Production Agric.* 6(4): 492–500

Uusitalo, R., Yli-Halla, M. and Turtola, E. (2000) Suspended soil as a source of potentially bioavailable phosphorus in surface runoff waters from clay soils. *Water Res.* 34: 2477–2482

Components of root exudates from genetically modified cotton

Weiming Shi and Weidong Yan

Institute of Soil Science, Chinese Academy of Sciences, Nanjing 210008, China

Abstract

The cultivation of genetically modified (GM) crops is hotly debated around the world, but the area of these crops has increased rapidly in the past 5 years. One of the major GM crops in China is transgenic (Bt) cotton, about half of which is produced by domestic biotechnologists and the remainder is imported. China ranks fourth in the world in the total area of cotton grown. The interactions between GM crops and soils through the root-soil interface are particularly important. To study these interactions, we assessed the root exudate components from Bt cotton. The growth medium's pH decreased from 6.0 to around 4 within 2 to 3 days, but no obvious difference was found between wild-type (WT) and Bt cotton under both normal and nutrient-reduced conditions. Acidification around the roots was also identified in agar media, with stronger acidification around mature roots and relatively weak acidification around root tips. Water-soluble components of root exudates were collected and the organic acids were analyzed by ion chromatography. Malic, oxalic, citric, and other organic acids were detected, with oxalic and citric acid the dominant species. The concentrations of organic acids in the root exudates were similar in both types of cotton. Further analysis of the root exudates is under way.

Keywords : genetically modified crops, transgenic cotton, root exudates, organic acids

1. Introduction

Cotton (*Gossypium hirsutum* L. [AD]$_1$) is a major Chinese crop. In the past two decades, the total cultivated area of cotton has ranged between 6.835 million ha in 1992 and 4.041 million ha in 2000. In 2001, 4.810 million ha of arable lands were used for cotton production. During the same period, total cotton production ranged from 3540 kt in 1976 to 5680 kt in 1991; in 2001 it equaled 5320 kt, at an average yield of 1.1 ton/hm^2. Of this total, between 300?000 and 600?000 ha represented genetically modified (GM) cotton. The major variety of GM cotton was transgenic cotton containing a modified version of the Bt (*Bacillus thuringiensis*) toxin (Cry1A or CpTI). Although this total area remains small compared with the area of non-transgenic varieties, the rate of increase in the adoption of GM cotton is very high. Transgenic cotton was first released in a field experiment in 1997, in a 667-hm^2 area, and was fully released into the open environment the following year. In 1999, the area planted with GM cotton increased to 148?000 hm^2. As of 2002, China ranked fourth in the world in terms of the total area of cotton. The cultivation of GM crops is still hotly debated in China (Meng, 1996; Fan et al., 2001) as well as elsewhere in the world (Donegan et al., 1995; Escher et al., 2000; Wolfenbarger and Phifer, 2000), but the actual area cultivated with GM cotton in China is expected to increase rapidly in the coming years.

Ecological assessments are performed before GM crops are released into the open environment and agricultural fields in accordance with China's regulations on GM crops. However, the current regulations that govern these assessments are loose, and we believe that more careful environmental assessments are necessary, particularly relating to the interactions between GM crops and soils through the root-soil interface. Therefore, as the first step in a more rigorous assessment of these crops, we examined the components of root exudates using transgenic Bt cotton as our study material. These root exudates are the predominant components that affect soil-plant interactions. Some reports have shown that root exudates influence the activities of soil microorganisms (Heuer et al., 2002; Saxena et al., 1999; Wang et al., 2002). In the present paper, we also report the effects of the nutritional status of the growth medium on changes in the components of cotton root exudates. This paper reports our preliminary results.

2. Materials and methods

2.1. Plant materials

The cotton cultivar studied in our trial was Sumian 12, a transgenic cotton purchased from the seed company with the modified Cry1A transgene inserted into transformed cotton with an otherwise identical genetic background.

2.2. Medium pH based on in situ observations

Both regular (WT) and transgenic (Bt) cotton seeds were germinated in quartz sand watered with tap water. After the cotyledon appeared, a complete nutrient solution was supplied and the plants were pre-cultured in a growth chamber (Sanyo) under a light period of 14 hours at 25 ℃ and 100 μE m^{-2} s^{-1} and a dark period of 10 hours at 23 ℃; relative humidity was maintained at 70% in both treatments. To test the effects of low-nutrient conditions, 10-day-old seedlings were pretreated with a nutrient-deficient solution for 2 days. The control seedlings were supplied with the aforementioned complete nutrient solution. Each plant was carefully transplanted into 1.0% (w/v) agar medium and placed under lighting. The agar medium was adjusted to pH 6.0, and contained 1 mM CaSO$_4$ and 60 mg L^{-1} bromophenol purple. Certain times later, the acidification of the agar around the roots was recorded photographically as changes in the color of the indicator dye, as described by Liu et al. (1997).

2.3. Change in the pH of the nutrient solution during plant growth

Two similar seedlings that had been pre-cultured as described above were transplanted into separate 500-mL pots that were then supplied with either the complete nutrient solution or the nutrient-deficient solution described above. The solution pH was adjusted to 6.0 from the beginning, and the solution pH was measured daily using a pH meter.

2.4. Collection and analysis of root exudates

The 10-day-old seedlings (x per treatment) were transferred into 500-mL pots containing either the complete or the nutrient-deficient solutions and grown for 1 week. Root-washings were collected by placing six plants from each treatment in 900 mL of de-ionized water for 3 h (between 9:00 am and 12:00 noon), and were then freeze-dried. The root exudates were dissolved in ultra-pure water and adjusted to a final volume of 0.5 mL.

The concentrations of organic acids were measured using a Shimadzu LC-6A high-performance liquid chromatograph (HPLC) with a Bondapak C-18 column. The effluent solution was 0.5% (v/v) phosphoric acid, and detection of the organic acids was performed using a UV spectrometer at a wavelength of 214 nm.

2.5. Plant analysis

The nitrogen, phosphorus, and potassium contents of the plant materials were analyzed using the standard methods described by Lu (2000). Acid-phosphatase activity was measured using the method of Zhou (1987).

3. Results and discussion

3.1. Change in the pH of the nutrient solution during plant growth

Changes in the medium pH and in the quantity of secreted protons were observed over the course of the growing period. Medium pH decreased rapidly, from its starting point at pH 6.0 to a value of around 4 within 2 to 3 days. However, no obvious difference was found between the WT and Bt cotton in either nutrient solution (Fig. 1). The area of acidification around the roots was also identified in the agar medium: the strongest acidification occurred around mature roots, whereas acidification around the root tips was relatively weak (Fig. 2). The photographs in Figure 2 indicate that although the ability to acidify the growing medium differed significantly between the two nutrient solutions, the introduction of the foreign *Cry1A* gene into the Bt cotton did not affect the acidification process or intensity around the roots.

3.2. Changes in nutrient compositions in cotton shoots

Unlike the other parameters we studied, the nutrient content of the cotton shoots differed significantly between the Bt and WT cotton, and was affected by the nutrient solution used. In the complete nutrient solution, the Bt cotton took up and accumulated more N, P, and K in its shoots than did the WT cotton (Fig. 3). In the N-deficient medium, levels of N in Bt cotton shoots decreased rapidly to levels below those in WT cotton, whereas levels of P and

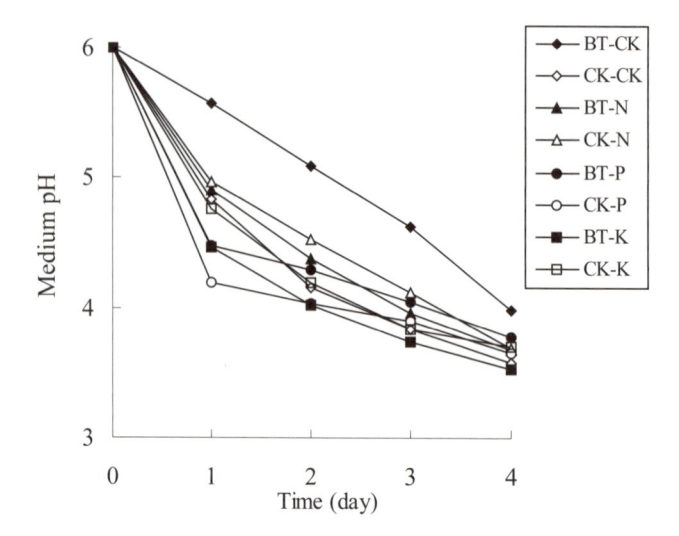

Fig. 1 Changes of medium pH following the time in WT and Bt-transgenic cotton under various nutrients supply status

BT-CK: Bt-transgenic cotton with full nutrients solution, CK-CK: WT cotton with full nutrients solution; BT-N: Bt-transgenic cotton with minus N nutrient solution, CK-N: WT cotton with minus N nutrient solution; BT-P: Bt-transgenic cotton with minus P nutrient solution, CK-P: WT cotton with minus P nutrient solution; BT-K: Bt-transgenic cotton with minus K nutrient solution, CK-K: WT cotton with minus K nutrient solution. The plants are 2-week-old. A pH meter recorded the medium pH.

Fig. 2 Local acidification of medium around roots of WT and Bt-transgenic cotton under various nutrients supply status

BT-CK: Bt-transgenic cotton with full nutrients solution, CK-CK: WT cotton with full nutrients solution; BT-N: Bt-transgenic cotton with minus N nutrient solution, CK-N: WT cotton with minus N nutrient solution; BT-P: Bt-transgenic cotton with minus P nutrient solution, CK-P: WT cotton with minus P nutrient solution; BT-K: Bt-transgenic cotton with minus K nutrient solution, CK-K: WT cotton with minus K nutrient solution. Plants are 10-d old. Medium is 1.0% agar with the pH indicator of bromophenol purple. The yellow region indicates the local acidification by roots secreted proton.

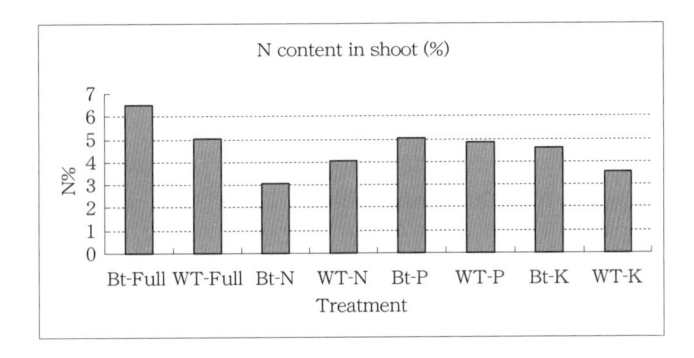

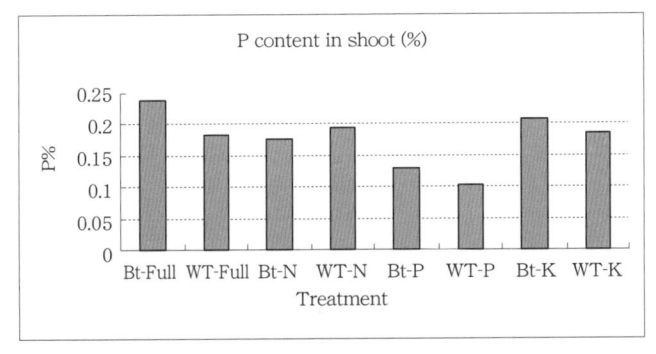

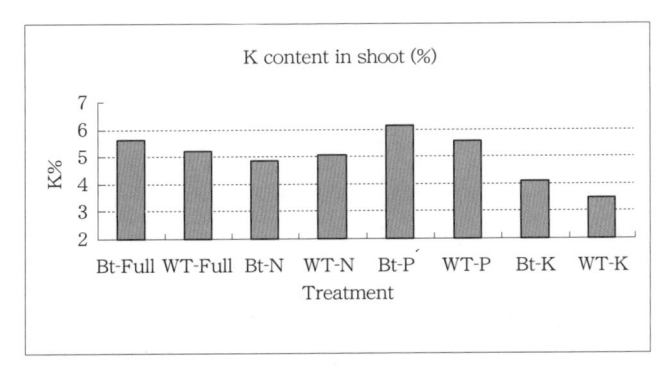

Fig. 3 Changes of N, P and K content in the shoot part of WT and Bt-transgenic cotton under various nutrients supply conditions

Bt-Full: Bt-transgenic cotton with full nutrients solution, WT-Full: WT cotton with full nutrients solution; Bt-N: Bt-transgenic cotton with minus N nutrient solution, WT-N: WT cotton with minus N nutrient solution; Bt-P: Bt-transgenic cotton with minus P nutrient solution, WT-P: WT cotton with minus P nutrient solution; Bt-K: Bt-transgenic cotton with minus K nutrient solution, WT-K: WT cotton with minus K nutrient solution. Ten-day old plants were treated for one week and harvested for analysis. N content was analyzed by Kjeldahl method, P content by colorimetry and K content by flame spectrometry.

K did not differ significantly. In the K-deficient medium, levels of N were much higher in Bt cotton than in WT cotton. We do not fully understand this phenomenon, though one possible explanation is that Bt cotton has a superior ability to take up K, and when K was removed from the nutrient solution, the Bt cotton was still able to take up N, probably in the form of ammonium ions. Further study is necessary to confirm this hypothesis.

3.3. Comparison of enzyme activity in leaves and roots

It has been reported that transgenic cotton that expresses Bt crystal proteins shows different patterns for certain important isoenzymes such as peroxidases compared with WT cotton (Ding et al., 2001). For this reason, we also measured the changes in acid-phosphatase activity between Bt and WT cotton under various nutritional conditions. The results (Fig. 4) showed that acid-phosphatase activity in both roots and leaves decreased sharply in N-deficient solutions. Neither P nor K deficiency appeared to increase acid-phosphatase activity in either type of cotton, and the two cotton types did not differ significantly in their response. Although many reports have indicated that short-term P deficiency increases acid-phosphatase activity, long-term deficiency will decrease acid-phosphatase activity. In the present

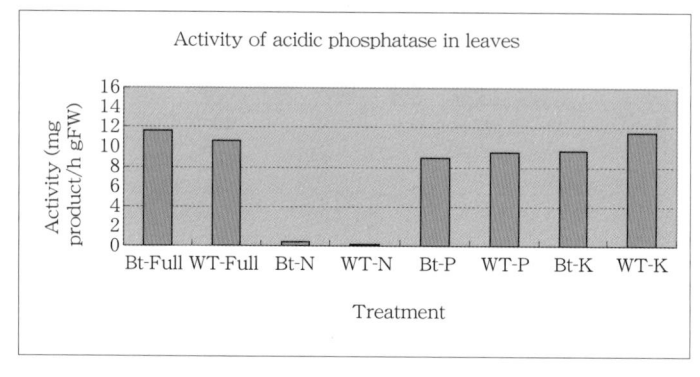

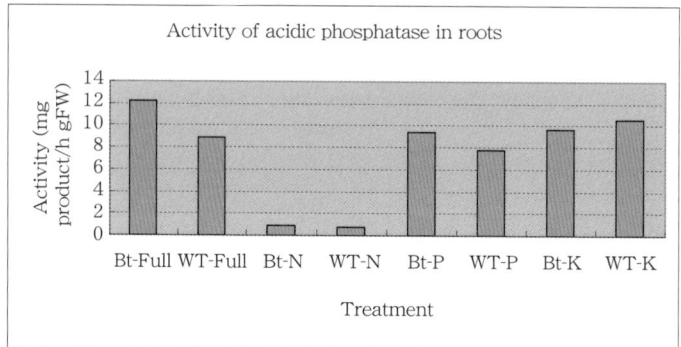

Fig. 4 Changes of acidic phosphatase in leaves and roots of WT and Bt-transgenic cotton under various nutrients supply conditions

Bt-Full: Bt-transgenic cotton with full nutrients solution, WT-Full: WT cotton with full nutrients solution; Bt-N: Bt-transgenic cotton with minus N nutrient solution, WT-N: WT cotton with minus N nutrient solution; Bt-P: Bt-transgenic cotton with minus P nutrient solution, WT-P: WT cotton with minus P nutrient solution; Bt-K: Bt-transgenic cotton with minus K nutrient solution, WT-K: WT cotton with minus K nutrient solution. Ten-day old plants were treated for one week and harvested for analysis. Enzyme activity was measured according to Reference Zhou LK.

study, our investigation lasted only 7 days, so we could not see phosphatase going up which may have occurred at the early period of treatment. The total acid-phosphatase activity was similar for Bt and WT cotton in both roots and leaves.

3.4. Changes in the composition of organic acids in root exudates

The water-soluble components of root exudates were collected from root-washings from WT and Bt cotton, and the organic acids fraction was analyzed by means of HPLC. The results are summarized in Table 1 and Figure 5. Malic, oxalic, citric, and seven other organic acids were detected in these fractions, with oxalic acid the dominant species (Table 1), followed by malic acid and citric acid. Our preliminary data showed that the total concentration of organic acids in the root exudates of Bt and WT cotton were not significantly different. However, the concentrations of the different organic acid species varied between nutrient treatments and differed between WT and Bt cotton. For example, tartaric acid was present at considerably higher levels in Bt cotton, whereas malonic acid was found at considerably higher levels in the root exudates of WT cotton. Our analysis of other components of the root exudates, such as sugars and amino acids, is currently under way, and will be reported in a future paper.

Over the past decade, considerable attention has been paid to the ecological risks of GM crops to the environment; these include gene flow from a transgenic species into other species, and the safety of GM food to humans. The impact of the root exudates from transgenic plants on the soil ecosystem has been considered to be a relatively low concern. However, it has been demonstrated that the Bt insecticidal toxin could be released from the roots of transgenic plants into the rhizosphere, and in some cases the toxin could persist for long periods. However, our results concerning organic acids did not support the hypothesis that the components of root exudates would differ significantly between transgenic Bt cotton and WT cotton, even though the nutritional status of these two types of cotton clearly differed under different nutrient conditions. These differences should be studied in more depth to confirm and expand upon our preliminary findings.

Table 1 Organic acids detected in root exudates of regular cotton and Bt cotton

Treatment	Oxalic acid	Tartaric acid	Formic acid	Pyruvic acid	Malic acid	Manolic acid	Ketoglutaric acid	Acettic acid	Citric acid	Succinic acid	Total amount
Bt-Full	5644	447	693	10	749		48	426	138	529	8684
WT-Full	5201		657	41	309	837	23	505	369		7942
Bt-N	3288		527	5	313	99	54	282	498	12	5078
WT-N	3238	190	391	7	393		18	383	527	3	5150
Bt-P	5037	585	306	23	189		27	644	585	155	7551
WT-P	5030		145	6	252	607	16	149	406	22	6633
Bt-K	3685	709	545	8	65	320	63	282	534		6211
WT-K	4300		484	10	299	418	60	156	53	130	5910

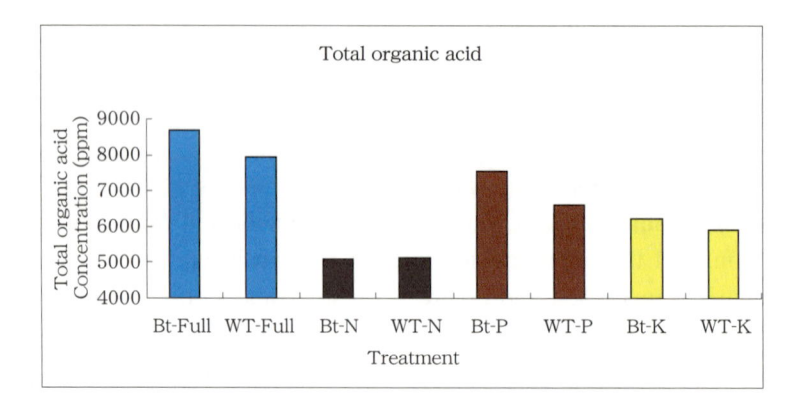

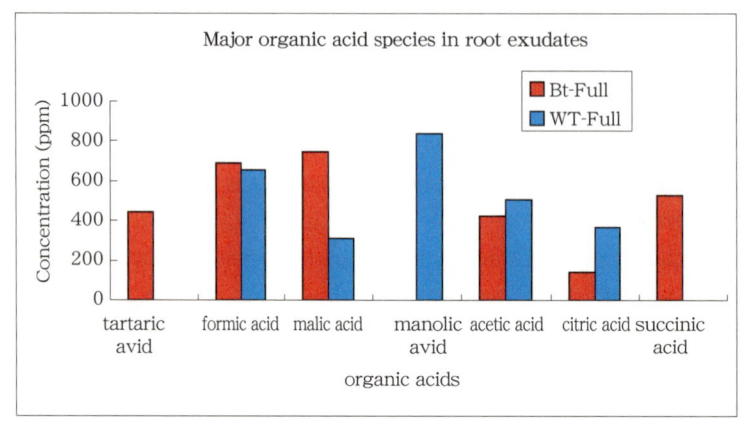

Fig. 5 Changes of total organic acid concentration and individual organic acid species in root exudates of WT and Bt-transgenic cotton under various nutrients supply conditions
Bt-Full: Bt-transgenic cotton with full nutrients solution, WT-Full: WT cotton with full nutrients solution; Bt-N: Bt-transgenic cotton with minus N nutrient solution, WT-N: WT cotton with minus N nutrient solution; Bt-P: Bt-transgenic cotton with minus P nutrient solution, WT-P: WT cotton with minus P nutrient solution; Bt-K: Bt-transgenic cotton with minus K nutrient solution, WT-K: WT cotton with minus K nutrient solution. Ten-day old plants were treated for one week and harvested for analysis. Organic acid was detected by HPLC.

Acknowledgment

This project was partially supported by a grant from the Institute of Soil Science of the Chinese Academy of Sciences.

References

Ding, Z.Y., Xu, C.R. and Wang, R.J. (2001) Comparison of several important isoenzymes between Bt cotton and regular cotton. *Acta Ecologica Sinica* 21: 332–336

Donegan, K.K., Palm, C.J. and Fieland, V.J. (1995) Changes in levels, species, and DNA fingerprints of soil microorganisms associated with cotton expressing the *Bacillus thuringiensis* var *kurstaki* endotoxin. *Applied Soil Ecology* 2: 111–124

Escher, N., Kach, B. and Nentwig, W. (2000) Decomposition of transgenic *Bacillus thuringiensis* maize by microorganisms and woodlice *Porcellio scaber* (Crustacea: Isopoda). Basic Applied Ecology 1: 161–169

Fan, L., Zhou, X., Hu, B., Shi, C. and Wu, J. (2001) Gene dispersal risk of transgenic plants. *Chinese Journal of Applied Ecology* 12: 630–632

Heuer, H., Kroppenstedt, R.M., Lottmann, J., Berg, G. and Smalla, K. (2002) Effects of T4 lysozyme release from transgenic potato roots on bacterial rhizosphere communities are negligible relative to natural factors. *Applied and Environmental Microbiology* 68: 1325–1335

Liu, Z., Li, L. and Shi, W. ed. (1997) Research Methods in Rhizosphere. Jiangsu Science and Technology Publishing House, Nanjing.

Lu, R. ed. (2000) Analytical Methods for Soils and Plants. China Agricultural Science Press, Beijing.

Meng, K.Q. (1996) Debating for biosafety of genetically engineered plants. *Progress in Biotechnology* 16: 2–6

Saxena, D., Florest, S. and Stotzky, G. (1999) Insecticidal toxin in root exudates from Bt corn. *Nature* 402: 480

Wang, Z.H., Ye, Q.F., Shu, Q.Y., Cui, H.R., Xia, Y.W. and Zhou, M.Y. (2002) Impact of root exudates from transgenic plants on soil micro-ecosystems. *Chinese Journal of Applied Ecology* 13: 373–375

Wolfenbarger, L.L. and Phifer, P.R. (2000) The ecological risks and benefits of genetically engineered plants. *Science* 290: 2088–2093

Zhou, L. ed. (1987) Soil Enzymology. Science Press, Beijing.

Methods for assessing the indirect effects of introduced hymenopteran parasitoids on Japanese agricultural ecosystems

Atsushi Mochizuki

National Institute for Agro-Environmental Sciences, 3-1-3 Kannondai, Tsukuba, Ibaraki 305-8604, Japan

Abstract

The introduction of natural enemies has been used to control exotic pests. Recently, there has been much discussion of the risk of indirect effects on local communities and on non-target species caused by populations of exotic natural enemies. In 1999, Japan's Ministry of the Environment released a report ("Guidelines on Introduction of Environmental Impact Assessment of Biological Control Agents") designed to help prevent undesirable environmental impacts. However, the methods for assessing these environmental impacts have not yet been determined. This paper describes techniques that may meet this need, based on laboratory experiments, DNA markers, and published and unpublished data on the classical biological control of insects.

Keywords: biological control, environmental impact assessment, indirect effects, introduced hymenopteran parasitoids

1. Introduction

Biological control is a key component of crop protection around the world. Classical biological control, which involves the introduction of exotic biological control agents to permanently suppress exotic pests, has been practiced for many years in many countries. In Japan, *Rodalia cardinalis* (Coleoptera: Coccinellidae) was first imported from Taiwan to control *Iserya purchasi* (Homoptera: Coccidae) in 1911. Satisfactory pest control was achieved by this introduction. After that success, many other natural enemies have been introduced from foreign countries, including hymenopteran parasitoids (Table 1). As the demand for organic farming products increases, farming technology based on the use of biological control agents has attracted public attention because of the risk of potential new environmental concerns. Compared with chemicals and antibiotics, biological control agents

Table 1 Introduced hymenopteran parasitoids in Japan

Parasitoid wasps	Origin	Year	Target pests	Results
Classical introduction				
Scutellista cyanea	America	1924	Red wax scale	failed
Prospaltella smithi	China	1925	Spiny blackfly	succeeded
Cryptognatha sp.	China	1925	Spiny blackfly	failed
Spathius fuscipennis	Philippines	1928	Rice stem borer	failed
Trichogramma chilonis	Philippines	1929	Rice stem borer	failed
Uscana semifumipennis	Hawaii	1930 − 31	Bean weevils	failed
Aphelinus mali	France	1926 − 27	Woolly apple aphid	failed
	America	1931		succeeded
Aneristus ceroplastae	Hawaii	1932	Red wax scale	failed
Opius fletcheri	Taiwan	1932	Melon fly	?
Macrocentrus ancylivorus	America	1933	Oriental fruit moth	failed
Glypta rufiscutellaris	America	1934	Oriental fruit moth	failed
Aphytis lingnanensis	America	1955	Arrowhead scale	failed
	Hong Kong	1972 − 76		succeeded
Aphytis yanonensis	China	1980	Arrowhead scale	succeeded
Cccobius (=Physcus) fulvus	China	1980	Arrowhead scale	succeeded
Copidosoma desantisi	Chile	1956	Potato tubermoth	failed
Copidosoma koehleri	India	1966	Potato tubermoth	?
Torymus sinensis	China	1975 −	Oriental chestnut gall wasp	succeeded
Olesicampe benefactor	Canada	1984	Larch sawfly	?
Bathyplectes curculionis	America	1988 − 89	Alfalfa weevil	continued
Microctonus aethiopoides	America	1988 − 89	Alfalfa weevil	continued
Commercial introduction				
Encarsia formosa	Netherlands	1995*	Whiteflies	continued
Eretmocerus eremicus	America	2002	Whiteflies	continued
Diglyphus isaea	Netherlands	1997	Leafminers	continued
Dacnusa sibirica	Netherlands	1997	Leafminers	continued
Aphidius colemani	Netherlands	1998	Aphids	continued

* introduced as an inoculative agent from England in 1975. added to Murakami (1997)

are considered to be safer from the viewpoint of human health. However, despite this advantage, it is possible that biological control agents, and especially agents introduced from foreign countries, may disturb agricultural and other ecosystems; these consequences arise because the biocontrol organisms can reproduce and migrate. There has been much discussion of the risk of indirect effects on local communities and non-target species posed by populations of exotic natural enemies (e.g., Howarth, 1983, 1991; Simberloff and Stiling, 1996).

At the end of the 20th century, the introduction of commercial biological control agents such as biotic pesticides have increased in Japan at the expense of classical methods. To help prevent undesirable impacts on the environment, Japan's Ministry of the Environment released a report in 1999, "Guidelines on Introduction of Environmental Impact Assessment of Biological Control Agents". In this paper, I discuss the Japanese guidelines and methods for assessing the potential indirect effects caused by exotic hymenopteran parasitoids in Japanese ecosystems.

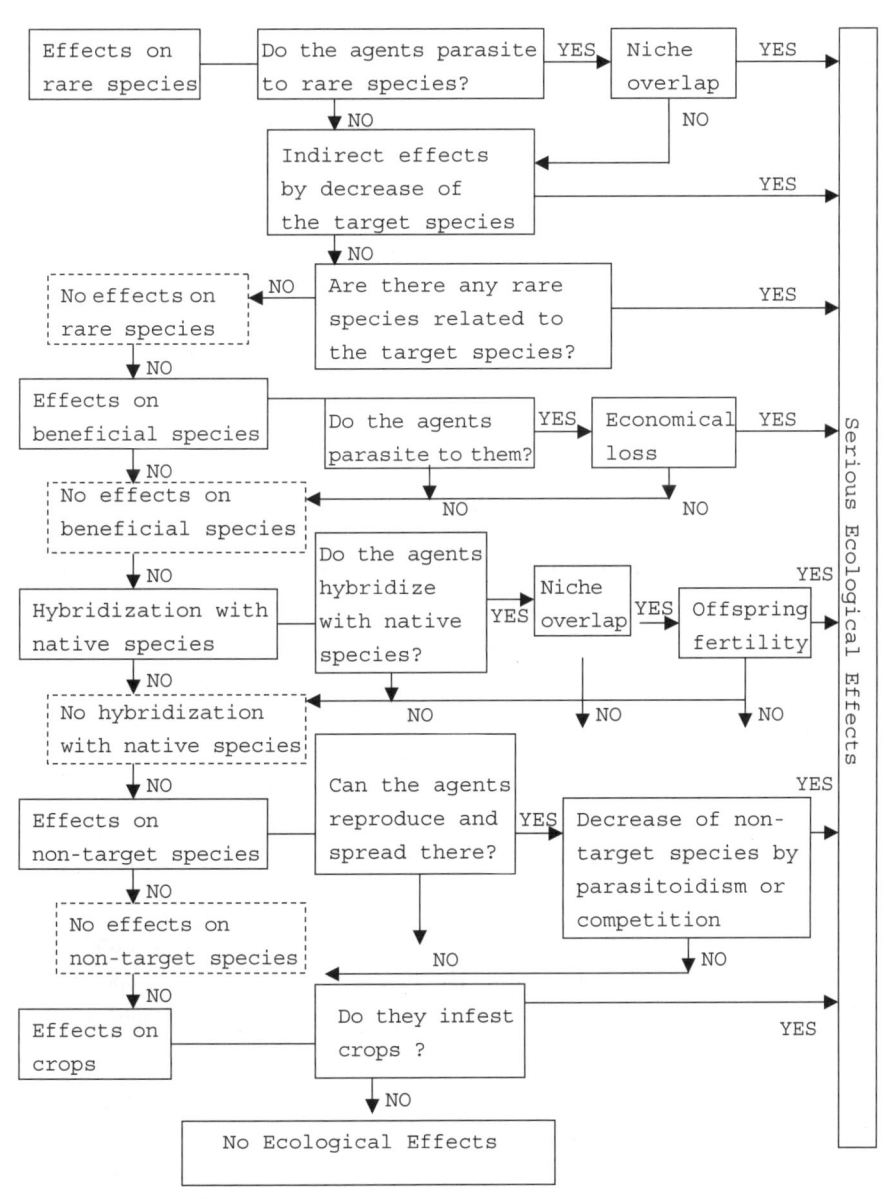

Fig. 1 Flow Chart for Assessment of Environmental Effects of Biological Control Agents (From "Guidelines on Introduction of Environmental Impact Assessment of Biological Control Agents")

2. Assessment items and method

2.1 Japanese guidelines

In the Japanese guidelines, the following items must be examined to detect any impacts caused by introduced natural enemies (Fig. 1):

   i. Effects on rare insects
   ii. Effects on beneficial insects (e.g., the silk worm, honeybees)
   iii. Hybridization between introduced and indigenous parasitoids
   iv. Direct or indirect effects on non-target insects (e.g., keystone species, symbol species, native parasitoids)
   v. Infestation of crops

Item (i) is difficult to study because most rare insects are difficult to collect or monitor in the field. Items (ii) to (iv) can be studied in a laboratory. Laboratory data often produce critical evaluations. However, it will be necessary to prepare for the highest risk. For parasitoid wasps, item (v) can be neglected because all these wasps are carnivorous.

2.2 Methods for improving assessments

The host range and the potentials for hybridization and competition are important factors that must be studied to address these five items in the guidelines. Laboratory experiments must be direct methods to know the host range but they cannot do so for all species. The host range of a parasitoid may sometimes be predicted by using data on close phylogenic relatives. For example, Belshaw et al. (1998) reconstructed the phylogeny of braconid and ichneumonid parasitoids on the basis of the variation of partial sequences of a 28S nuclear rRNA gene (rDNA). In this study, they overlapped the life history of each genus to create a phylogenetic tree. Related species have similar biological and life history traits.

The life history of parasitoid wasps are classified into two main categories: *Idiobionts* are wasps that kill their host immediately before oviposition. Most idiobiont species lay their eggs outside the host, and hatched larvae feed on the dead host. In contrast, *koinobionts* let the larval host continue developing more or less normally following the parasitization event. Most koinobiont species lay their eggs inside their host, and the hatched larvae grow inside their host without killing it before pupation. The host range of koinobiont species is typically narrower because they have evolved much more precise behavioral and immunological coordination with the biology of their hosts (Quicke, 1997).

The possibility of hybridization between introduced and indigenous parasitoids is also related to their phylogenic positions. Offspring between two species can be detected by using DNA markers for various nuclear genes. An F_1 female will have two sets of the nuclear genes, one set from each parent. If some nuclear gene sequences differ between the parents, the offspring will have two *different* sets of genes. Internal transcribed spacers (ITS) of rDNA are useful for discrimination.

In terms of the interactions between introduced and native parasitoids, there are three main types of competitive interaction: between idiobiont species, between idiobiont and koinobiont species, and between koinobiont species. Laboratory experiments were performed with parasitoid wasps of the bean leafminer, *Liriomyza torifolii*. First, competition between the introduced idiobiont (*Diglyphus isaea*) and two native idiobionts (*Hemiptarsenus varicornis* and *Neochrisocharis formosa*) was studied. Leafminer larvae were provided as hosts for oviposition by two parasitoid species at a time, and the species that emerged from the parasitized larvae were recorded. The oviposition intervals between the two parasitoid species were 1 hour, 1 day, and 3 days. Post-ovipositional species always emerged, regardless of introduce or native species.

One competition study compared an idiobiont with an introduced koinobiont (*Dacnusa sibirica*). Koinobiont species are usually less competitive than idiobiont species because idiobionts kill their host before oviposition. However, koinobiont species can avoid competition with idiobionts by adopting the following strategies:

(1) Inhibition of superparasitism : Some koinobiont species guard their offspring against parasitism by idiobionts through the use of chemical deterrents (e.g., Gates, 1993).

(2) Niche shifting to avoid direct interactions: Koinobiont species can typically oviposit on any stage of the host (e.g., Shimada, 1985), but idiobionts prefer more mature hosts (Fig. 2). Some koinobiont species oviposit more than idiobionts on more mature hosts in the presence of idiobionts.

Whether *D. sibirica* can avoid competition by adopting these strategies was studied. The first strategy was rejected because the idiobionts *D. isaea* and *H. varicornis* oviposited on the host regardless of the presence or absence of parasitism by *D. sibirica*. In the second test, *D. sibirica* preferred to oviposit on more mature hosts in the presence of the sympatric idiobiont *D. isaea*, but oviposited irrespective of the host stage in the presence of an allopatric idiobiont, *H. varicornis*. The results of these experiments suggest the following predicted effects:

(1) Introduced (exotic) idiobiont parasitoid wasps will not influence parasitism by native

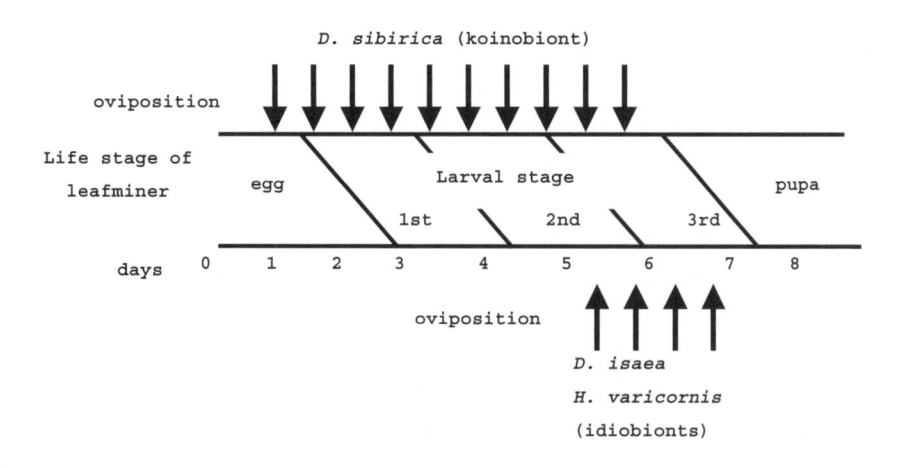

Fig. 2 Parasitization schedule of parasitic wasps

idiobiont species.

(2) Introduced (exotic) idiobiont parasitoid wasps may influence parasitism by native koinobiont species. The introduced idiobiont parasitoids may even exterminate indigenous koinobionts in closed systems such as greenhouses.

The accumulation of experience from previous (classical) biological control programmes is also important. Lynch et al. (2001) reviewed existing published and unpublished data on classical biological control of insects. In Japan, many parasitoid wasps have been introduced and released, but only six species have successfully controlled pests. Only two parasitoids, *Aphytis yanonensis* (for *Unaspis yanonensis*) and *Torymus sinensis* (for *Dryocosmus kuriphilus*), have thus far been studied to detect population fluctuations and relationships with native parasitoids after the initial interaction. However, the indirect effects of the introduction remain unknown.

3. Conclusions

Exotic biological control agents will affect native ecosystems to a greater or lesser extent. It is likely that in the near future, risk-benefit analyses will be required for all prospective biological control agents, and releases will be permitted only when the benefits of an introduction are expected to significantly outweigh the risks to native and beneficial species. Based on the research summaries in this paper, the following assessment tools should be considered as a means for predicting the impacts of introductions:

1. Close phylogenic relatives of the introduced agent should be studied as an aid in predicting the effects of the introduction. Genetic markers can be used to establish these relationships, as well as to predict the likelihood of hybridization.
2. Competition between introduced and native idiobionts and koinobionts should be studied in the laboratory.

References

Belshaw, R., Fitton, M., Herniou, E., Gimeno, C. and Quicke, D.L.J. (1998) A phylogenic reconstruction of the Ichneumonoidea (Hymenoptera) based on the D2 variable region of 28S ribosomal RNA. *Systematic Entomology* 23: 109–123

Gates, S. (1993) Self and conspecific superparasitism by the solitary parasitoid Antrocephalus pandens. Ecological Entomology 18: 303–309

Howarth, F.G. (1983) Classical biocontrol: panacea of Pandora's box? Proceedings of the Hawaiian Entomological Society 24: 239–244

Howarth, F.G. (1991) Environmental impacts of classical biological control. *Annual Review of Entomology* 36: 485–509

Lynch, L.D., Hokkanen, H.M.T., Babendreier, D., Bigler, F., Burgio, G., Gao, Z-H., Kuske, S., Loomans, A., Menzler-Hokkanen, I., Thomas, M.B., Tommasini, G., Waage, J.K., wan Lenteren, J.C. and Zeng, Q-Q. (2001) Insect biological control and non-target effects: a European perspective. In Wajnberg, E., Scott, J.K. and Quimby, P.C. (eds) Evaluating Indirect Ecological Effects of Biological Control. CABI Publishing, Wallingford, UK. : 99–125

Murakami, Y. (1997) Natural Enemies of the Chestnut Gall Wasp-Approaches to Biological Control. (in Japanese). Kyushu University Press, Fukuoka, Japan: 308 pp

Quicke, D.L.J. (1997) Parasitic Wasps. Chapman & Hall, London, UK: 470 pp

Shimada, M. (1985) Niche modification and stability of competitive systems. II. Persistence of interspecific competitive systems with parasitoid wasps. *Researches on Population Ecology* 27: 203-216

Simberloff, D. and Stiling, P. (1996) How risky is biological control? *Ecology* 77: 1965-1974

Monitoring of environmental resources and their utilization in South Korea

Mun-Hwan Koh, Goo-Bok Jung and Ki-Cheol Eom

National Institute of Agriculture Science and Technology, Suwon 441-707, Korea

Abstract

Monitoring the chemical properties of agricultural soils permits the development of improved soil-management (fertilization etc.) systems and remediation practices for polluted soils. In the present study, we monitored soil fertility and heavy metal contents in benchmark soils of paddy fields, uplands, greenhouses, and orchards, as well as soils near wastewater treatment plants, closed metal mines, an industrial complex, and a highway area. Soil pH, organic matter, available phosphate, extractable K, Ca, and Mg, and heavy metals (Cd, Cu, Pb, Zn, As, Ni, Cr) were analyzed. Average contents of organic matter, available phosphate, and extractable K increased more rapidly in the greenhouse's soils than in upland or paddy soils, but farmers clearly overfertilized their crops in all three areas. In agricultural land, heavy metal contents were below the threshold values that defined soil contamination, but in areas near the abovementioned industrial and urban areas, levels were often high enough to require rehabilitation. To rehabilitate soils polluted with heavy metals, we recommend the application of a thick layer of topsoil, land reconsolidation, and soil amelioration with lime, phosphate, organic matter, or flooding, plus the cultivation and harvesting of inedible crops such as trees.

Keywords : monitoring soils, fertility, heavy metals, arable land, fertilizer, recommendations

1. Introduction

South Korea has implemented an intensive agricultural system to support the country's large population from a small cultivated area, and has pursued an agricultural policy based on high yield by applying relatively large amounts of fertilizer per hectare. As a result, the country has achieved self-sufficiency in rice production by increasing the yield per hectare throughout the 1990s, aided by the development of a domestic inorganic fertilizer industry since the 1960s. Unfortunately, salinity has increased in some agricultural lands owing to the continuous application of inorganic fertilizers. In some major vegetable cultivation areas and greenhouses, plant growth has been inhibited by the accumulation of salts and nutrients that has resulted from excessively high inputs of inorganic and organic fertilizers (Jo et al., 1999; Park et al., 2001).

Recently, the need for environmentally friendly sustainable agriculture has created a need to minimize the inputs of agricultural materials and pesticides so as to avoid environmental problems. Therefore, new fertilizer application techniques must be developed to protect the environment and produce food safely; the goal is to save the country's natural resources by applying appropriate levels of fertilizers to every crop (NIAST, 1999, 1999-2002). However, farmers have tended to maintain their practice of intensive fertilization to ensure high crop productivity. As a result, problems have arisen in safe crop production and in environmental contamination arising from the imbalance between nutrient inputs and crop nutrient demands, resulting in an excessive accumulation of nutrients in agricultural lands (Kim, 1990; NIAST, 1970-2000; Park et al., 2001).

Establishment of an environmentally friendly sustainable agricultural system has been made possible by a comprehensive survey in support of the implementation of a sustainable agricultural policy in 1998 (RDA, 2001). This initiative led to the establishment of a 4-yearly survey to monitor soil resources and agricultural environments. To provide the required data for controlling fertilizer application and reclaiming land, the accumulation of nutrients in soils, soil chemical properties, and heavy metals were monitored in rice paddies, uplands, greenhouses, and orchards (NIAST, 1999-2002). Community problems have arisen related to the safety of crop production and to crop injury resulting from heavy metal contamination from industrial complexes and metal mines. The degree of heavy metal contamination has also been monitored in agricultural lands near highways, an industrial complex, and closed metal mines, as well as near the inflow area for municipal wastewater around a city with suspected environmental contamination (Jung et al., 2002; Oh, 1997; Ryu et al., 1995; Yun et al., 2002). Rehabilitation methods for soils contaminated with heavy metals and proper fertilizer application levels in agricultural lands were proposed on the basis of the national monitoring results.

The objective of this paper is to summarize the fundamental data required to implement environmentally friendly sustainable agriculture based on proper management guidelines for soil rehabilitation and fertilization for every crop. These data have been gathered by comprehensive monitoring of soil chemical properties and heavy metals in South Korean agricultural land since the 1960s.

2. Materials and Methods

2.1. Fertilization systems and soil chemical properties in agricultural land

The population and area of agricultural lands used in this paper were obtained from annual statistical reports produced by the South Korean government (MAF, 2000; RDA, 2000). Monitoring of soil chemical properties in agricultural lands has produced comprehensive data for different cultivation types, such as rice paddies, uplands, plastic greenhouses, and orchards for the past 40 years. The range of proper chemical properties for soils was evaluated by using monitoring data gathered since 1999. These data were determined by using standard soil chemical analysis methods of the Rural Development Administration (Jo et al., 1999; NIAST, 1970-2000, 1999-2002, 2000). Data for fertilizer application rates and methods for different crops and cultivation types were described by focusing on the results of previous research and the annual research reports of the Rural Development Administration (NIAST, 1999, 1970-2000; Park et al., 2001).

2.2. Countermeasures for rehabilitating soils contaminated with heavy metals and a survey of heavy metal contents in agricultural lands

To obtain data on levels of heavy metals in agricultural lands, researchers have monitored rice paddies, uplands, greenhouses, and orchards since 1999 and have compared them with agricultural land near municipal wastewater treatment plants, closed metal mines, an industrial complex, and a highway (Jung et al., 2002; NIAST, 1999-2002). Heavy metal contents in the soil were determined by using the standard analysis methods for contaminated soils used by South Korea's Environmental Protection Agency (Ministry of the Environment, 2001). Data on the rehabilitation of soils contaminated with heavy metals were obtained from previous research papers and from the annual research reports of the Rural Development Administration (Jung et al., 2000; Kim, 1990; Kim et al., 1985; NIAST, 1970-2000; Yun et al., 2002).

3. Results and Discussion

3.1. Land use in South Korea

The total land area of South Korea equals 99 810 km^2, representing 45% of the Korean peninsula. This area is relatively small compared with other countries in the region. In 2000, the area of cultivated land in South Korea equaled 18?890 km2, or 19% of the total land area (Table 1). The area of agricultural land has decreased rapidly since the 1980s, but rice paddies still accounted for 60.8% of agricultural land in 2000. The utilization rate of agricultural land decreased from 142.1% in 1970s to 110.5% in 2000. In 2000, the mean

Table 1 Trends of land use in Korea

Items	Unit	'70	'80	'90	'95	'00
Total land (A)	1,000 ha	9,848	9,899	9,927	9,927	9,981
Cultivated land (B)	1,000 ha	2,298	2,196	2,109	1,985	1,889
- Paddy field	1,000 ha	1,273	1,307	1,345	1,206	1,149
- Upland	1,000 ha	1,025	889	764	779	740
Utilized land (C)	1,000 ha	3,264	2,765	2,409	2,197	2,098
- B/A	%	23.2	22.2	21.2	20.0	18.9
- C/B	%	142.1	125.3	113.3	110.7	110.5
Cultivated land per farm household	ha	0.92	1.02	1.19	1.32	1.37
- Paddy field	ha	0.51	0.61	0.76	0.80	0.83
- Upland	ha	0.41	0.41	0.43	0.52	0.54

Table 2 Current status of the Korea agriculture

Items	'70	'80	'90	'95	'00
GDP (Billion Won, A)	2,771	36,857	178,797	377,350	482,744
Agricultural (B)	646	5,612	13,027	20,042	20,350
Ratio (B/A, %)	23.3	15.2	7.3	5.3	4.6
Farmer's population ratio (%)	44.7	28.4	15.5	10.9	8.6

cultivated area per household was 0.83 ha of rice paddy and 0.54 ha of upland.

The total cultivated area decreased by an average of 23 333 ha annually during the 1990s. However, arable land expanded continuously by around 1860 ha/year through the reclamation of mountainous areas and coastal areas (RDA, 2000). This increase could not replace the 27 760 ha/year that was converted to buildings, public facilities, and other urbanization purposes. The fact that more than 1% of the total arable land per year is being converted into non-agricultural uses represents a serious problem.

The rapid industrialization in South Korea that began in the early 1960s has caused a continuous decline in the contribution of the agricultural sector to GDP and in the proportion of the population involved in farming (Table 2). Agriculture's share of GDP decreased from 23.3% in 1970 to 15.2% in 1980 and to 4.6% in 2000 (MAF, 2000). The agricultural sector's share of total employment has also declined, from 44.7% in 1970 to 15.5% in 1990 and 8.6% in 2000. Moreover, the national percentage of farm employment has been decreasing at a similar pace.

3.2. Characterization of soil chemical properties

Periodic monitoring of the chemical properties of agricultural soils is urgently required before it can become possible to establish appropriate fertilizer application regimes and a scientific approach to soil reclamation in an intensive cultivation system such as the one that has developed in South Korea. Recently, a system for monitoring the chemical properties of agricultural soils has been established to create benchmarks based on the area, topography, and cultivation type across the nation (Table 3). The enactment of a policy for environmentally friendly sustainable agriculture by the Rural Development Administration in 1998 established a comprehensive monitoring system to detect changes in the agricultural environment, and in the establishment of benchmarking of soils every 4 years for each cultivation

Table 3 Monitoring of chemical properties in agricultural Fields

Paddy	Plastic film house	Upland	Orchard
'95 (1,168)*	'96 (513)	'97 (854)	'98 (547)
'99 (4,047)	'00 (2,651)	'01 (1,650)	'02 (1,360)
'03 (2,010)	'04 (1,260)	'05 (1,650)	'06 (1,360)

*Year (number of samples)

type across the nation (Jo et al., 1999; NIAST, 1999-2002; RDA, 2001). The objective of this initiative was to provide the fundamental data required for the development of environmentally friendly sustainable agriculture through proper soil management and proper application of fertilizer in each crop.

Available phosphorus contents have increased continuously in rice paddies since 1960, but the other nutrients that were monitored have not shown a consistent trend (Table 4). Average contents of available phosphorus and exchangeable K were higher than optimal in 1999, but

Table 4 Changes of chemical properties in paddy soils (adopted from NIAST, 1999)

Period	pH (1:5)	OM (g kg^{-1})	Av·P$_2$O$_5$ (mg kg^{-1})	Av·SiO$_2$ (mg kg^{-1})	Ex. Cations (cmol+kg^{-1})			Samples
					K	Ca	Mg	
'64-68	5.5	26	60	78	0.23	4.5	1.8	5,130
'76-'79	5.9	24	88	75	0.31	4.2	1.3	19,737
'80-'87	5.7	23	107	88	0.27	3.8	1.4	616,687
'90	5.7	27	101	80	0.32	4.3	1.5	1,192
'95	5.6	25	128	72	0.32	4.0	1.2	1,168
'99	5.7	22	136	86	0.32	4.0	1.4	4,047
Optimum	6.0-6.5	25-30	80-120	130-180	0.25-0.30	5.0-6.0	1.5-2.0	

Table 5 Distribution ratio of chemical properties in paddy soils in year 1999 (adopted from NIAST, 1999)

Classification		pH (1:5)	OM (g kg^{-1})	Av·P$_2$O$_5$ (mg kg^{-1})	Av·SiO$_2$ (mg kg^{-1})	Ex. Cations (cmol + k g^{-1})		
						K	Ca	Mg
Average		5.7	22	136	86	0.32	4.0	1.4
Optimum range		6.0-6.5	25-30	80-120	130-180	0.25-0.30	5.0-6.0	1.5-2.0
Dist. (%)	Lower	80.3	48.4	38.3	90.1	45.4	80.7	72.8
	Opt.	14.9	24.6	21.3	6.2	19.7	13.5	14.7
	Excess	4.8	27.0	40.4	3.7	34.9	5.8	12.5

Table 6 Changes of chemical properties in upland soils (adopted from NIAST, 2001)

Period	pH (1:5)	OM (g kg^{-1})	Av·P$_2$O$_5$ (mg kg^{-1})	Ex. Cations (cmol + kg^{-1})			Samples
				K	Ca	Mg	
'64-'68	5.7	20	114	0.32	4.2	1.2	3,661
'76-'80	5.9	20	195	0.47	5.0	1.9	18,324
'85-'88	5.8	19	231	0.59	4.6	1.4	65,565
'92	5.5	24	538	0.64	4.2	1.3	854
'97	5.6	24	577	0.80	4.5	1.4	854
'01	5.9	24	547	0.81	5.8	1.6	1,650
Optimum	6.0-6.5	20-30	300-500	0.50-0.60	5.0-6.0	1.5-2.0	

the contents of other nutrients were suboptimal. The proportion of samples in which soil properties were lower than optimum ranged from 38.3% to 90.1% (Table 5). In particular, pH, available SiO_2 and exchangeable Ca were lower than optimal by 19.7%, 9.9%, and 19.3%, respectively. These findings suggest that it will be necessary to apply silicate-slag and magnesium-superphosphate fertilizer in rice paddies.

In upland soils, organic matter, available phosphorus, and exchangeable Ca increased consistently since 1960, and had especially dramatic increases after 1990 (Table 6). All soil chemical properties in Table 6 appeared to be near-optimal or higher than optimal in 2001. In the upland soils, the proportion of soil chemical properties that were lower than optimal ranged from 25.1% to 58.1% (Table 7). Therefore, it appears that soil fertility was slightly better than in rice paddies. Available phosphorus and exchangeable K exceeded optimum levels in 54.1% and 53%, respectively, of the samples. This may have resulted from the practice of applying fertilizers without considering the phosphorus and K contents of the organic fertilizers that were also used, and from excessive application of both inorganic and organic fertilizers to increase crop yields, particularly in vegetable cultivation. We recommend that fertilizer applications be reduced, because farmers are clearly applying greater amounts of fertilizer than are required.

There was a tendency for values of all soil chemical properties to increase in greenhouses (Table 8). Levels of these properties were generally higher than optimal, with the exception

Table 7 Distribution ratio of chemical properties in upland soils in year 2001 (adopted from NIAST, 2001)

Classification		pH (1:5)	OM (g kg^{-1})	Av・P_2O_5 (mg kg^{-1})	Ex. Cations (cmol + kg^{-1})		
					K	Ca	Mg
Average		5.9	24	547	0.81	5.8	1.6
Optimum range		6.0-6.5	20-30	300-500	0.50-0.60	5.0-6.0	1.5-2.0
Dist. (%)	Lower	58.1	56.6	25.1	36.7	50.6	45.6
	Opt.	20.9	31.6	20.7	10.3	12.9	12.9
	Excess	21.0	11.7	54.1	53.0	36.5	41.5

Table 8 Changes of chemical properties in soils under vinyl house (adopted from NIAST, 2000)

Period	pH (1:5)	EC (dS m^{-1})	OM (g kg^{-1})	Av・P_2O_5 (mg kg^{-1})	Ex. Cations (cmol + kg^{-1})			Samples
					K	Ca	Mg	
'64-'68	5.8	–	22	811	1.08	6.0	2.5	215
'76-'79	5.8	3.7	26	945	1.01	6.4	2.3	391
'91-'93	6.0	1.9	31	861	1.07	5.9	1.9	1,072
'94-'95	6.1	2.7	30	876	1.11	6.5	2.2	216
'96	6.0	2.8	35	1,092	1.27	6.0	2.5	513
'00	6.3	2.8	34	975	1.67	7.7	3.4	2,651
Optimum	6.0-6.5	2.0>	20-30	300-500	0.70-0.80	5.0-6.0	1.5-2.0	

of pH; in particular, available phosphorus and exchangeable K and Mg have been roughly twice the optimal values since 1990. The proportion of greenhouse samples in which soil properties were suboptimal ranged from 9.5% to 32.3% in 2000 (Table 9). Because the proportions of samples in which properties exceeded optimal levels was so high, it appears that greenhouse soils were overfertilized compared with rice paddies and uplands. Available phosphorus and exchangeable K were particularly excessive, with 79.1% and 72.0%, respectively, of samples exceeding optimal levels. This may have resulted from excessive application of organic and inorganic fertilizers because of the higher yield sought and population density used in vegetable culture in greenhouses. The accumulation of nutrients in these soils caused growth and salt injuries to crops as a result of nutrient imbalances. Therefore, soil rehabilitation and fertilizer application adjusted to account for actual soil chemical properties (based on testing of soils before planting) might be necessary to solve secondary problems with water contamination (due to leaching of minerals) and crop injury in greenhouses.

In orchards, the average soil nutrient contents tended to increase, but organic matter decreased and available phosphorus decreased after an overall increase during the 1990s (Table 10). The contents of available phosphorus and exchangeable K were nearly double the optimum levels. The proportion of samples with suboptimal soil chemical properties ranged from 17.8% to 56.7% (Table 11). This suggests that the status of soil fertility in orchards was similar to that in uplands. However, available phosphorus and exchangeable K were higher than optimal in 65.8% and 58.9%, respectively, of samples. This tendency was similar to

Table 9 Distribution ratio of chemical properties in soils under vinyl house in year 2000 (adopted from NIAST, 2001)

Classification		pH (1:5)	OM (g kg^{-1})	Av・P$_2$O$_5$ (mg kg^{-1})	Ex. Cations (cmol + kg^{-1})		
					K	Ca	Mg
Average		6.3	34	975	1.67	7.7	3.4
Optimum range		6.0-6.5	20-30	300-500	0.70-0.80	5.0-6.0	1.5-2.0
Dist. (%)	Lower	32.3	30.4	12.3	22.9	21.6	9.5
	Opt.	53.2	29.5	8.5	5.0	25.0	29.3
	Excess	14.5	40.1	79.1	72.0	53.4	61.2

Table 10 Changes of chemical properties in orchard soils (adopted from NIAST, 2002)

Period	pH (1:5)	OM (g kg^{-1})	Av・P$_2$O$_5$ (mg kg^{-1})	Ex. Cations (cmol + kg^{-1})			Samples
				K	Ca	Mg	
'93	6.0	22	444	0.75	4.9	1.3	608
'94	5.5	29	762	0.83	5.1	1.4	357
'98	5.7	29	780	0.82	5.8	1.5	507
'02	5.9	23	588	0.96	5.8	1.7	1,360
Optimum	6.0-6.5	25-30	200-300	0.30-0.60	5.0-6.0	1.2-2.0	

Table 11 Distribution ratio of chemical properties in orchard soils in year 2002 (adopted from NIAST, 2001)

Classification		pH (1:5)	OM (g kg^{-1})	Av·P$_2$O$_5$ (mg kg^{-1})	Ex. Cations (cmol + kg^{-1})		
					K	Ca	Mg
Average		5.9	23	588	0.96	5.8	1.7
Optimum range		6.0–6.5	25–30	200–300	0.30–0.60	5.0–6.0	1.2–2.0
Dist. (%)	Lower	56.5	54.9	23.2	17.8	52.4	56.7
	Opt.	20.4	23.0	11.0	23.3	11.5	18.3
	Excess	23.1	22.1	65.8	58.9	36.2	25.0

those in uplands and greenhouses.

3.3. Best fertilization practices

In recent years, it has become necessary to modify farming practices so that farming will become a sustainable industry that produces safe agricultural products. The use of various chemical compounds, such as pesticides and fertilizers, and the raising of livestock in cities have both adversely affected the environment. Farmers cultivating cash crops have usually applied too much fertilizer to increase yields, because fertilizer is cheap in South Korea. Farmers can choose between two main recommended fertilizer application rates: the national standard and a level calculated based on soil test data (NIAST, 1999). Recommended standard fertilization levels have been established for 77 crops and readjusted for 59 crops, and equations have been established to calculate fertilizer application rates for 70 crops. These standard levels are also varied to account for differences in topography and soil type (for example, depending on the texture and drainage class of paddy soils).

By monitoring soil fertility, it is possible to adjust the standard fertilizer application rates based on the average values measured by soil tests for various crops (Table 12). In the results described in section 3.2, K and phosphorus contents in continuously cultivated soils have steadily increased, so the adjusted fertilization levels use reduced amounts of these nutrients.

Recommended fertilizer application rates based on soil analyses have been provided to

Table 12 Readjustment of fertilizer recommendation based on the changes of nationwide soil fertility (adopted from NIAST, 1999)

Crops	Existing (kg ha^{-1})			Readjusted (kg ha^{-1})		
	N	P$_2$O$_5$	K$_2$O	N	P$_2$O$_5$	K$_2$O
House melon	250	77	160	187	63	109
House cucumber	240	164	238	197	103	122
Melon	250	200	250	250	77	160
Cucumber	300	200	300	240	164	238
Rice	110	70	80	110	45	57

Table 13 Models for fertilizer recommendation based on soil test for paddy rice (adopted from NIAST, 1999)

Fertilizer	Equation to calculate the application amount of fertilizers
N	kg N / 10 a = 12.74 − 1.52 OM + 0.028 SiO$_2$
P$_2$O$_5$	kg P$_2$O$_5$ / 10 a = (100 − P$_2$O$_5$) × 0.1
K$_2$O	kg K$_2$O / 10 a = (0.03 × CEC − K) × 47.1
SiO$_2$	kg SiO$_2$ − Fertilizer / 10 a = (130 − SiO$_2$) × 3.84
Fine earth	ton Soil / 10 a = (18 × (15 − Clay content)/(25 − Clay content)) × 12
Organic	kg Compost / 10 a = 2000 (OM < 2.0 %)
	kg Compost / 10 a = 1500 (2.0 ≤ OM ≤ 3.0 %)
	kg Compost / 10 a = 1000 (OM ≥ 3.0 %)
Zinc	kg Zn / 10 a = 3 (pH ≥ 7.0 and SiO$_2$ − Fertilizer ≥ 300 kg / 10 a)
Gypsum	kg CaSO$_4$ / 10 a = 51.6 × CEC − 86 × Ca / (Ca / CEC < 0.6 and pH ≥ 7.0 and Saline soil)

farmers through a national network. Equations have been developed to calculate the recommended application rate based on the results of simple soil tests (Table 13). For example, the nitrogen application rate is based on the soil's contents of organic matter and SiO$_2$, whereas applications of phosphorus, K, SiO$_2$, and organic matter are based on their respective contents in the soil. To keep inputs low and to protect the environment, the system for allocating fertilizer to farmers is now based on soil tests conducted by each village, which demonstrate the actual nutrient needs of local crops (MAF, 1999).

Over the past 50 years, the recommended fertilizer application rate increased dramatically until the 1990s, and was based largely on the new crop varieties being cultivated, on a farmer's desire and income, and on soil fertility (Table 14). However, application rates began

Table 14 Changes of fertilizer recommendation rate of NPK for crops during the last 50 years (adopted from NIAST, 1999). (Unit : kg ha^{-1})

Crop	Fertilizers	1950s	1960s	1970s	1980s	1990s
Rice	N	35	67	117	110	110
	P$_2$O$_5$	38	53	58	70	48
	K$_2$O	34	60	65	80	68
Cereals	N	−	94	99	110	90
	P$_2$O$_5$	−	72	87	109	44
	K$_2$O	−	67	76	92	44
Vegetables	N	−	−	−	259	251
	P$_2$O$_5$	−	−	−	187	104
	K$_2$O	−	−	−	249	180

decreasing in the 1990s in response to soils test data.

The amount of fertilizer used in rice paddies increased until the 1980s, then decreased in the 1990s. This can be partially attributed to excessive application of fertilizer as a result of the use of manure compost made by farmers, which exceeded the use of inorganic fertilizers until the 1970s. The trends for rice paddies are similar to those for uplands. Recommended amounts of fertilizers were apparently reduced because phosphorus and calcium accumulated to excessive levels since the 1990s in uplands and where vegetable culture has been practiced.

3.4. Changes in heavy metal contents, and countermeasures for contaminated soils

The threshold levels for soil contamination and corrective actions were established through the enactment of a law on soil and environment conservation in 1996. Pollutants regulated under this law included eight heavy metal species and 16 species of organic component, including organic P, phenole, and trichloroethylene (Table 15). The criteria for soil contamination distinguish two groups according to the nature of the land: "A" areas include those around agricultural land, forestry areas, schools, streams, water services, and parks. "B" areas include those around industrial areas, factories, and railroads. The regulatory criteria set lower thresholds for soil contamination in "A" areas and define levels of corrective action based on the degree of soil contamination.

The "threshold level for soil contamination" is the criterion used to regulate land use and the installation of new facilities so as to avoid serious damage to plant, animal, and human health as a result of soil contamination. The "corrective action level" is the criterion that dictates when rehabilitation is required and is designed to prevent the contamination of soils

Table 15 Threshold and corrective action values for heavy metals contamination of soils in Korea (adopted from MAF, 2001). (Unit : mg kg^{-1})

Pollutant	Threshold level		corrective action level	
	A sites	B sites	A sites	B sites
Cd	1.5	12	4	30
Cu	50	200	125	500
As	6	20	15	50
Hg	4	16	10	40
Pb	100	400	300	1,000
Cr^{6+}	4	12	10	30
Zn	300	800	700	2,000
Ni	40	160	100	400
F	400	800	800	2,000
Organic-P	10	30	–	–
Cyanamide	2	120	5	300
Phenole	4	20	10	50
Trichloroethylene (ECE)	8	40	20	100

A : Agricultural Land, Forestry, School, Stream, Water Service, Park, Sports Area.
B : Industrial, Factory, Railroad, and Others Area.

to levels of more than 40% of the threshold level. The evaluation criteria for heavy metal contamination in a soil are based on the soluble concentrations of Cd, Cu, Pb, As, and Cr and on the total concentrations of Hg, Zn, and Ni.

Surveys of the degree of contamination of agricultural lands began in the late 1970s with investigations of closed metal mines by the Rural Development Administration. Heavy metals were also investigated in uncontaminated agricultural lands with different culture types, such as paddy fields, uplands, greenhouses, and orchards, starting in the early 1990s. The heavy metal contents were also monitored in agricultural lands near highways, near industrial complexes, and near closed metal mines. Rehabilitation of contaminated soils was based on the monitoring results.

With the enactment of a policy for environmentally friendly sustainable agriculture by the Rural Development Administration in 1998, a comprehensive system for monitoring changes in the agricultural environment was established, and the benchmarking of soils to detect these changes was performed every 4 years for each cultivation type and pollutant (Table 16). Changes in heavy metals in agricultural materials were investigated through the monitoring system, and provided the fundamental data used to direct soil amendments on lands contaminated with heavy metals.

In order to evaluate the heavy metal contents of soils in all agricultural land in South Korea, 4047 paddy soils, 2651 greenhouse soils, 1650 upland soils, and 1360 orchard soils were analyzed, with the results classified by topography and type of culture (Table 17). Levels of Zn were relatively high in greenhouses, but there was little difference between types of culture, and no levels exceeded the thresholds. This observation may have resulted from the application of livestock manure compost containing high concentrations of Zn, from excessive application rates, and from the poor quality of available manure compost (Jo et al., 1999; Park et al., 2001). The heavy metal contents of uncontaminated agricultural land were generally low and below the safety threshold based on the criteria for soil contamination described in South Korea's Soil Environmental Conservation Act (Ministry of the Environment, 2001).

Table 16 Monitoring of heavy metal contents in agricultural
(adopted from Jo et al., 1999; Jung et al., 2002)

Lands	Non-contaminated agricultural fields			
	Paddy	Plastic film house	Upland	Orchard
Year (samples)	'95 (1,168) '99 (4,047) '03 (2,010)	'96 (513) '00 (2,651) '04 (1,260)	'97 (854) '01 (1,650) '05 (1,650)	'98 (547) '02 (1,360) '06 (1,360)
Lands	Unfavorable agricultural fields			
	Urban area	Metal mine area	Industrial area	Highway area
Year (samples)	'99 (600) '03 (600)	'00 (600) '04 (600)	'01 (600) '05 (600)	'02 (600) '06 (600)

Table 17 Average contents of heavy metals in agricultural fields in 1999–2002 (adopted from NIAST, 1999–2002). (Unit : mg kg^{-1})

Fields	Cd	Cu	Pb	As	Zn	Cr	Ni
Paddy	0.105	4.70	4.84	0.59	4.5	0.34	0.67
Plastic film house	0.117	4.82	2.69	0.36	31.2	0.69	1.19
Upland	0.078	3.17	2.54	0.42	10.6	0.73	2.55
Orchard	0.084	3.04	2.20	0.66	10.1	0.38	1.00
Threshold level	1.5	50	100	6	300	4	40

Table 18 Average contents of heavy metals in the unfavorable agricultural fields for environmental contamination (adopted from Jung et al., 2002). (Unit : mg kg^{-1})

Fields	Cd	Cu	Pb	As	Zn	Cr	Ni
Urban area*	0.136	7.31	7.45	0.50	11.1	0.36	1.26
Metal mine	0.586	17.88	22.61	3.68	34.6	0.26	0.99
Industrial	0.283	9.16	9.03	1.02	8.3	0.45	0.92
Highway	0.097	4.04	5.20	0.46	4.7	0.17	0.59
Threshold level	1.5	50	100	6	300	4	40

*Flowing area of discharge water from municipal wastewater treatment plant.

In contaminated agricultural lands, heavy metal concentrations were highest near closed metal mines, followed by industrial complexes, areas where waters are discharged from municipal wastewater treatment plants, and highways (Table 18). The average contents of heavy metals in these areas were typically 400% to 600% the levels in paddy soils. However, heavy metal contents near discharge areas for waters from municipal wastewater treatment plants and in agricultural lands near highways were similar to those of uncontaminated land. But agricultural land near closed metal mines and an industrial complex exceeded the criteria for soil contamination (Ministry of the Environment, 2001).

The chemical species that exceeded thresholds most often were Cd, Cu, Pb, and As (Table 19). The numbers of sites that exceeded soil contamination thresholds and corrective action levels were 57 (9.5%) and 120 (20%), respectively. Furthermore, the frequency of exceeding these levels was greatest for As, followed by Cd, Cu, and Pb. The numbers of sites exceeding the contamination and corrective action thresholds in agricultural lands near industrial complexes were each 17 (5.7%).

Heavy metal contents exceeded threshold levels at 20% of 36 metal mine areas (120 sites), and exceeded corrective action levels at 9.5% of these areas (57 sites). Heavy metal contents exceeded threshold levels at 2.8% of 5 industry complex areas (17 sites), and exceeded corrective action levels at 2.8% of these areas (17 sites). These percentages translate into 204.3 and 84.1 ha near closed metal mines that exceeded the threshold and corrective action levels, respectively, for 36 sites of 58 investigated locations (Table 20). For industrial

Table 19 Numbers of sites exceeding the threshold and corrective action values for soil contamination in the unfavorable agricultural fields (adopted from NIAST, 1999–2002)

Fields	Samples	Cd	Cu	Pb	As	Sum
Urban area	600	0	0	0	0	0
Mine area	600	54 (7)*	38 (13)	25 (3)	61 (40)	120 (57)
Industrial area	600	5 (9)	10 (6)	2 (0)	10 (5)	17 (17)
Highway area	600	0	0	0	0	0
Total	2,400	59 (16)	48 (19)	27 (3)	71 (45)	137 (74)

* Exceeded site numbers to the criteria for soil contamination with threshold and corrective action values.

Table 20 Areas exceeding the threshold and corrective action values for soil contamination in the unfavorable agricultural fields (adopted from NIAST, 1999–2002)

Fields	Year	Exceeded area	Areas exceeding soil contamination criteria (ha)		Sum
			Threshold	Corrective action	
Mine area	'00	36	204.3	84.1	288.4
Industrial area	'01	5	181.2	42.1	223.3
Total		41	385.5	126.2	511.7

complexes, the corresponding percentages translate into 181.2 and 42.1 ha that exceeded the threshold and corrective action values, respectively.

The results of investigating locations that exceeded the criteria for soil contamination were reported to South Korea's Environment and Agriculture & Forestry Ministries, and soil amendments were performed to correct the problems (NIAST, 1999–2002). The result of sampling sites after these treatments generally appeared to show an improvement in the degree of contamination in 2002. For polluted soils with levels of heavy metals above the threshold levels, the application of fine red earth, land preconsolation, flooding, and soil ameliorations such as the addition of lime, phosphate, and organic matter are recommended. For areas that require corrective action, the cultivation of non-edible crops such as trees, flowers, and fiber crops, as well as land reformation and the application of large amounts of fine red earth (up to 30 cm) are strongly recommended.

In South Korea, the contamination of paddy fields by heavy metals may have been caused by wastewater originating from abandoned metal mines and metal-processing industries. Generally, heavy metals have accumulated slowly but continuously in paddy fields and may have damaged crops directly (in the cases of Cu, Pb, Zn, and As) and may have damaged livestock and humans indirectly (in the cases of Cd and Hg) through the food chain (Berti and Jacobs, 1996; Fergusson, 1990; Krebs et al., 1998). One is occurred more serious toxicity than the other. Soil pH is considered to be the primary factor that dictates the effects of heavy metals in soils (Jung et al., 2000; Ryu et al., 1995). For example, these researchers found that uptake and translocation of Cd in plants could be decreased by increasing soil pH through the

Table 21 Changes of cadmium contents in brown rice as affected by slaked lime (adopted from Kim, 1990). (Unit: mg kg^{-1})

Lime application	Cd content in soil (mg kg^{-1})	
	1.51	5.40
No fertilizer	0.60	1.90
NPK	0.37	1.20
NPK + 150 kg/10a	0.26	0.96
NPK + 300 kg/10a	0.17	0.69

application of slaked lime in land contaminated with Cd. The Cd content in brown rice decreased significantly after the application of slaked lime with a high buffering capacity, and it was reduced even more by the addition of a higher quantity of slaked lime (Table 21).

In Cu-contaminated soils, rice yields decreased to 63.5% and 66.1% of control levels (under intermittent and continuous irrigation, respectively) at 200 mg Cu kg^{-1}, but the yield was increased to 79.9% and 80.1% of the control levels (respectively) with the application of slaked lime and to 93.9% and 84.0% of control levels (respectively) with the addition of wollastonite (Kim et al., 1985; Table 22). The improvement was greater with the addition of wollastonite under intermittent irrigation, but greater with the addition of slaked lime under continuous irrigation.

In order to select the most effective plants for use in the phytoremediation of areas polluted with heavy metals, Jung et al. (2002) cultivated eight species of inedible plants in soils near a metal smelter that had been contaminated with heavy metals, and analyzed the levels of absorbed heavy metals (Cd, Cu, Pb, and As) in each part of the plants. The plants included five trees (*Populus nigra* × *maximowiczii*, *Euonymus japonica*, *Acer palmatum*, *Celtis sinensis*, *Buxus microphylla*), two flowers (*Rhododendron lateritium*, *Calendula officinalis*), and a lawn (*Zoysia japonica*).

Growth by the trees was higher than that of the flowers and lawn. Heavy metals accumulated at higher levels in the roots than in the leaves and stems (Table 23). The concentrations of Cd and As were highest in the roots of *B. microphylla*. However, the total absorbed Cd was

Table 22 Effects of lime application on rice yield index in Cu treated soils (adopted from Kim et al., 1985)

Cu contents in soil (mg kg^{-1})	Intermittent irrigation			Continuous irrigation		
	Control	Slaked lime	Wollastonite	Control	Slaked lime	Wollastonite
0	100.0	—	—	100.0	—	—
50	101.0	103.6	106.6	101.2	104.4	104.6
100	94.5	98.8	99.1	98.1	102.0	100.7
200	63.5	79.9	93.9	66.1	80.1	84.0
Average	89.8	94.1	99.6	91.4	95.5	96.4

Table 23 Cd and As contents of testing plants cultivated at near a metal smelter (adopted from Jung et al., 2002)

Plants	Cadmium (mg kg^{-1})			Arsenic (mg kg^{-1})		
	Leaf	Stem	Root	Leaf	Stem	Root
P. nigra × P. m.	4.6	4.0	6.5	2.6	1.5	3.0
E. japonica	1.5	1.8	8.9	4.8	2.1	38.2
A. palmantum	2.5	3.0	7.9	11.8	1.5	35.9
C. sinensis	1.8	2.3	6.6	1.3	1.5	2.9
B. microphylla	2.4	3.4	18.1	6.7	5.9	242.8
R. lateritium	3.6	3.5	6.9	8.4	6.5	45.5
C. officinalis	7.9	15.8	11.6	6.7	5.8	2.5
Z. japonica	2.4	—	8.2	4.0	—	13.7

Table 24 Heavy metal uptake of testing plants cultivated at near a metal smelter (adopted from Jung et at., 2002).

Plants	Dry weight (kg/3.3 m^2)	Cd	Cu	Pb	As
		(mg/3.3 m^2)			
P. nigra × P. m.	144.4	644.2	1972.4	1699.7	273.8
E. japonica	18.1	63.8	555.9	352.9	212.1
A. palmantum	30.7	131.7	866.8	486.1	397.0
C. sinensis	83.7	261.6	2270.6	1302.5	146.0
B. microphylla	14.1	102.5	1319.2	403.0	1031.1
R. lateritium	6.1	28.8	290.2	124.7	127.5
C. officinalis	4.6	69.7	105.4	48.0	26.3
Z. japonica	13.2	67.6	534.4	410.8	112.2
Metal contents of experiment soil (mg kg^{-1})		1.93	102.0	141.1	61.4

highest for *P. nigra × maximowiczii*, followed by *C. sinensis*, *A. palmatum*, *C. sinensis*, and *B. microphylla*. Total accumulation of As was greatest in *B. microphylla*, followed by *A. palmatum* and *P. nigra × maximowiczii* (Table 24). *Populus nigra × maximowiczii*, *B. microphylla*, *A. palmatum*, and *C. sinensis* were thus the most effective species for phytoremediation of polluted areas based on the growth by these species and on their heavy metal uptake.

A 6-year analysis of heavy metal levels at two contaminated sites near a closed iron mine appears in Table 25. The contents of Cd, Pb, and Zn in agricultural lands near the closed iron mines decreased during the period of monitoring; Cu levels decreased in the final year of the assessment, but there was no clear annual trend. These results can be attributed to the addition of soil amendments, lime, and silicate-slag fertilizer by the Environment and Agriculture & Forestry Ministries (Ryu et al., 1995; Jung et al., 2000).

Table 25 Average content of heavy metal extracted by 0.1N–HCl in paddy soils near the closed metal mines (adopted from Yun et al., 2002) (Unit: mg kg^{-1})

Metal mine	Years	Cd	Cu	Pb	Zn
A mine (10)*	2001	2.92	36.1	97.2	114.1
	2000	4.15	41.3	126.4	148.2
	1995	5.93	38.8	212.6	185.6
B mine (20)	2001	0.61	4.9	11.9	33.7
	2000	0.59	5.6	12.7	42.3
	1995	1.20	4.2	17.7	70.6

* Number of sample

4. Conclusions

The enactment of a policy to support environmentally friendly sustainable agriculture by the Rural Development Administration in 1998 established a comprehensive monitoring system to detect changes in the agricultural environment, and particularly changes in the quality of soil resources. This monitoring provided the fundamental data required to permit science-based soil amendments and calculations of appropriate fertilizer application rates so as to prevent nutrient accumulation in soils. This initiative also provided data on appropriate soil amendments to produce safe agricultural products and to reduce crop damage due to overfertilization, as well as to protect ecosystems and humans from heavy metal contamination.

The annual development of new technologies by the Rural Development Administration gradually permitted a decrease in the number of chemical improvement studies required after 1985. However, the amount of research on manures and pollution increased during the 1990s. South Korea's research targets have also shifted from rice and grain crops to focus on vegetables, and from the amelioration of problems to precise management of nutrients so as to prevent the occurrence of problems.

Future soil management research goals in South Korea will focus on the development of optimum fertilization practices based on soil characteristics such as the soil's physical and morphological properties, fixations, and nutrient balances (release, input, and output). Moreover, research will lead to the development of "best practices" for land management using data obtained from field sensors that monitor variables such as chlorophyll levels, organic matter, electrical conductivity, levels of available K and P, moisture and drought, available soil depth, yield, crop density, growth status, and injury by insects and other pests, among other factors.

Agricultural land must be conserved in South Korea to produce high-quality food and maintain the country's ecological balance. Controlling the quantity and quality of waste materials created by human activities is a more important and practical strategy than rehabilitating soils after a problem arises.

References

Berti, W. R. and Jacobs, L. W. (1996) Chemistry and phytotoxicity of soil trace elements from repeated sewage sludge applications. *J. Environ. Qual* 25: 1025-1032

Brady, N. C. and Weil, R. R. (1996) The nature and properties of soils. Prentice Hall (Eleventh edition).

Fergusson, J. E. (1990) The heavy elements: Chemistry, environmental impact and health effects. Pergamon Press.

Jo, I. S., Jung, B. G. and Jung, G. B. (1999) A counter mersuring studies to the changes of agricultural environment. Project Research Report of National Institute of Agricultural Science and Technology, RDA.

Jung, G. B. (2001) A study on chemical fractions of heavy metal in soils and uptake by rice as affected by cultural practices. Ph. D. Thesis. Dankook University.

Jung, G. B., Kim, W. I., Moon, K. H. and Ryu, I. S. (2000) Extraction methods and availability for heavy metal in paddy soils near abandoned mining areas. *Kor. J. Environ. Agric.* 19 (4): 314-318

Jung, G. B., Kim, W. I., Lee, J. S. and Kim, K. M. (2002a) Studies on the phytoremediation of heavy metal contaminated soils by plants cultivation near a metal smelter. *Kor. J. Environ. Agric.* 21 (1): 31-37

Jung, G. B., Lee, S. B. and Lee, J. S. (2002b) Survey on the change of heavy metal contents and chemical properties in the vulnerable agricultural fields for environmental contamination. Project Research Report of National Institute of Agricultural Science and Technology, RDA.

Kim, B. Y. (1990) Soil contamination and countermeasurement. *Agro-engineering Technology.* 7 (2): 135-143

Kim, K. S., Kim, B. Y. and Han, K. H. (1985a) Studies on uptake of lead by crops and reduction of its damage. *J. Korean Soc. Soil Sci. Fert.* 19 (3): 217-221

Kim, K. S., Kim, B. Y., Lee, M. H., Han, K. H. and Kim, M. S. (1985b) Effect of water management and lime application on the growth and copper uptake of paddy rice. *Kor. J. Environ. Agric.* 4 (2): 102-107

Krebs, R., Gupta, S. K., Furrer, G. and Schulin, R. (1998) Solubility and uptake of metals with and without liming of sludge amended soils. *J. Environ. Qual.* 27: 18-23

MAF (1999) '99 Sustainable Agriculture Direct Income Support System. Ministry of Agriculture and Forestry.

MAF (2000) Statistical Yearbook of Agriculture and Forestry. Ministry of Agriculture and Forestry.

MOE (2001a) Soil Environmental Conservation Act. Ministry of the Environment.

MOE (2001b) Standard Test Method for Soil Pollution. Ministry of the Environment.

NIAST 1970-2000. Annual Report of Research project. Soil Management Division.

NIAST (1998) Strategy and Direction of Soil Management Research. National Institute of Agricultural Science and Technology, Soil Management Division.

NIAST (1999) Fertilizer Application Recommendation for Crops. National Institute of Agricultural Science and Technology, Sangrok press. Suwon.

NIAST 1999-2002. Annual report of the monitoring project on agro-environmental quality. National Institute of Agricultural Science and Technology

NIAST (2000) Methods of Soil and Plant Analysis. National Institute of Agricultural Science and Technology.

Oh, J. K. (1997) Evaluation of contamination at closed mine and application methods of tailing wastes. Symposium on the Remediation and Application Methods of Environmental Pollution around Abandoned Mine. 97-1. ILE. Forum of Environmental Policy: 15-51

Park, Y. H., Lee, J. Y. and Kim, F. J. (2001) Evaluation study on environmental affects by fertilizer use in cultivation lands. Project Research Report of National Institute of Agricultural Science and Technology, Rural Development Administration.

RDA (2000) Major Statistics of Korean Agriculture. Rural Development Administration.

RDA (2001) Environmental Friendly Agriculture Act. Rural Development Administration.

Ryu, S. H., Lee, J. R. and Kim, K. H. (1995) Sequential extraction of Cd, Zn, Cu, and Pb from the polluted paddy soils and their behavior. *J. Korean Soc. Soil Sci. Fert*. 28 (3): 207-217

Yun, S. G., Jung, G. B., Kim, W. I., Lee, J. S. and Kim, J. H. (2002) Impact assessment on the environmentally disturbed agricultural area. Agro-environmental Research 2001. National Institute of Agricultural Science and Technology, Rural Development Administration.

A 1:1 000 000-scale soil database and reference system for the People's Republic of China

Xuezheng Shi, Dongsheng Yu[a], Xianzhang Pan[a], Weixia Sun[a], Zitong Gong[a], Eric D. Warner[b] and Gary W. Petersen[b]

[a] *Institute of Soil Science, Chinese Academy of Sciences, Nanjing 210008, China*
[b] *Environmental Resources Research Institute, the Pennsylvania State University, University Park, USA*

Abstract

Soils maps of the People's Republic of China have been generated at different scales by means of ground surveys and laboratory analyses. The Officer for the Second National Soil Survey of China created a series of printed soil maps that cover the entire country at a scale of 1:1 000 000. The maps are now being converted into digital format. The digital database for these maps will consist of three parts: spatial soil data, soil attributes, and a reference system for Chinese soils. The spatial data is based on the "Soil Genetic Classification of China", and consists of 12 orders, 61 great groups, 235 sub-great groups, and 909 families. The soil maps are delineated on the basis of this classification. The sampled soil attributes include physical, chemical, and fertility properties for 2469 soil series. A comprehensive reference system will correlate the attributes for each soil series with names in three other soil classification systems: the "Soil Genetic Classification of China", the USDA Soil Taxonomy, and the FAO World Reference Base for soil resources. Cross-references will be provided via a relational database so that any researcher can learn the name of a soil in any of these systems.

Keywords : soil database, 1:1 000 000 soil maps, soil series of China, soil reference system, soil properties.

1. Introduction

A nationwide series of 1:1 000 000-scale soil maps for the People's Republic of China currently exists in traditional printed format. Conversion of these soil maps into digital form is a critical step that will let Chinese researchers and managers fully exploit the advantages that computer-based technologies provide, such as the ability to perform environmental and agronomic analyses using Geographic Information System (GIS) technology. This paper describes the assembly of the Chinese soil maps and concludes with an outline of the proposed structure for a digital version of the map series, which includes a tool for identifying the name of any soil in the database using three classification systems.

2. The Second National Soil Survey of China

The 1:1 000 000 soil maps of China were compiled from the Second National Soil Survey. This survey lasted 16 years, beginning in 1979 and finishing in 1994. The soils in 2444 counties, 312 national farms, and 44 forest farms were surveyed. Since 1990, the soil surveys have resulted in the publication of soil maps at scales ranging from 1:1 000 000 to 1:4 000 000 for the entire country. During the compilation of the soil maps, mapping units were delineated by using a genetic soil classification system. Through a nationwide collaboration by Chinese soil scientists, this classification system was initially conceived in 1953. The current form was finalized in 1978, then modified once more prior to its use in the Second National Soil Survey of China. The summary reports for the soil survey include "Soils of China" (Shi, 1998) and Volumes I – VI of the "Soil Series of China" (The Officer for the Second National Soil Survey of China, 1996). The soil surveys also supported many publications summarizing the information collected for other administrative units. In addition to the national summaries, soil maps, soil series, and descriptions of soil properties were published for each province. More than 5500 summary reports were prepared on the basis of the work done for the Second National Soil Survey of China.

3. Soil Classification System and Attributes of China's Soil Database

The "Soil Genetic Classification of China" is the standard system for delineating soil mapping units for the 1:1 000 000 soils map. The genetic system represents a four-level hierarchical structure, with 12 orders, 61 great groups, 235 sub-great groups, and 909 families in the database. Some of the 61 great groups are shown in Table 1, and five selected families of the Latosols sub-great group (from the Latosols great group) appear in Table 2.

China's soil database contains the spatial and other attributes of each soil. The spatial data are cited from the 1:1 000 000 soil maps, which have been compiled with basic map units representing soil families, whereas the soil attributes were derived from 2469 representative soil-series profiles included in Volumes I – VI of the "Soil Series of China". The sampled soil properties include physical properties (such as soil particle composition and soil texture), chemical properties (such as pH, organic matter content, cation exchange capacity, and exchangeable Ca, Mg, K, and Na), and fertility (such as total N, P, and K, and available P and K).

4. A Reference System for Chinese Soils

The development of the Chinese genetic classification system began in 1953, with modifications continuing until 1978, when the final structure was adopted. Since the genetic system became the foundation for national-level soil surveys, most soil data collected in China were analyzed on the basis of this system. However, the genetic classification is quite

Table 1 Some examples for soil orders and great groups for the 1:1,000,000 soils map

Orders	Great groups	Orders	Great groups
Ferrallisols	Latosols Latosolic red soils Red soils Yellow soils	Primarosols	Limestone soils Volcanic ash soils Purplish soils Phospho-calcic soils
Alfisols	Yellow-brown soils Yellow-cinnamon soils Brown soils Dark-brown soils Albic soils Brown coniferous forest soils Bleached podzolic soils Podzolic soils	Semi-aqueous soils Aqueous soils	Lithosols Skeletal soils Meadow soils Lime concretion black soils Mountain meadow soils Shrubby meadow soils Fluvo-aquic soils Bog soils
Semi-Alfisols	Torrid red soils Cinnamon soils Gray-cinnamon soils Black soils Gray-forest soils	Alkalin-saline soils	Peat soils Solonchaks Desert solonchaks Coastal solonchaks Acid sulphate soils
Pedocals	Chernozems Castanozems Castano-cinnamon soils Dark loessial soils	Anthrosols	Frigid plateau solonchaks Solonetzs Paddy soils Irrigation-warping soils
Aridisols	Brown caliche soils Sierozems	Alpine soils	Irrigated desert soils Felty soils
Desert soils	Gray desert soils Gray-brown desert soils Brown desert soils		Dark felty soils Frigid calcic soils Cold calcic soils
Primarosols	Loessial soils Red primitive soils Neo-alluvial soils Takyr Aeolian soils		Cold brown calcic soils Frigid desert soils Cold desert soils Frigid frozen soils

Table 2 Five selected soil families from the Latosols great group

Orders	Great groups	Sub-great groups	Family
Ferrallisols	Latosols	Latosols	Granitic Latosols Siliceous Latosols Red-stonic Latosols Argillitic Latosols Purplish Latosols

different from the "Chinese Soil Taxonomy", the United States Department of Agriculture (USDA) Soil Taxonomy, and the World Reference Base (WRB) of the United Nations' Food and Agriculture Organization (FAO). The differences between the systems present significant difficulties when Chinese researchers attempt to disseminate Chinese soils research and when international scientists attempt to understand Chinese soils research. To resolve this problem, the names of China's genetic soil families will be matched with those in the other classification systems, and the digital soil database will include a "reference system for Chinese soils" that translates between the systems. The correlation between the three different soil classification systems will be based on the soil properties obtained from the 2469 soil series

Table 3 The reference system for nine selected soil series

Location	Chinese genetic soil classification			Soil taxonomy	WRB
	Great group	Sub-great group	Soil Series		
Jinghong, Yunnan Province	Latosols	Latosols	Jinghong Latosols	Typic Hapludults	Alumic Acrisols
Jinghong, Yunnan Province			Jinghong clayed Latosols	Humic Hapludults	Alumic Acrisols
Yingjiag, Yunnan Province		Yellow Latosols	Yingjiang granitic Latosols	Typic Hapludults	Alumic Acrisols
Ruili, Yunnan Province	Latosolic red soils	Latosolic red soils	Sandy latosolic red soils	Ultic Hapludults	Dystric Luvisols
Lanchuan, Yunnan Province			Hongxiang latosolic red soil	Humic Rhoclicc Hapludox	Rhodic Ferralsols
Mojiang, Yunnan Province			Mojiang latosolic red soils	Typic Kandiudox	Rhodic Ferralsols
Lanchuan, Yunnan Province			Lanchuan clayed latosolic red soils	Typic Hapludults	Rhodic Acrisols
Menglian, Yunnan Province			Sandy latosolic red soils	Typic Rhodudults	Haplic Acrisols
Lanchuan, Yunnan Province			Lanchuan latosolic red soils	Mollic Paleudalfs	Chromic Luvisols

cited in Volumes I-VI of the "Soil Series of China", described previously. Table 3 lists the equivalent soil classification units in the three classification systems based on an interpretation of these soil-series data. The data associated with each soil series will include their equivalent names in three soil classification systems: the "Soil Genetic Classification of China", the USDA Soil Taxonomy (Soil Survey Staff, 1994), and the WRB Taxonomy (FAO, 1998). The cross-references between the three systems will be constructed in the form of a relational database so that any Chinese or international scientist can easily determine the name of a soil in any of the three systems.

5. Versions of China's soil database based on different soil-classification systems

Whether in international exchanges or in higher education, in China or elsewhere in the world, soil scientists and scientists in related fields are all eager to learn the distribution patterns and regional characteristics of the soils of China. To do so, they would prefer to use the USDA Soil Taxonomy or the WRB taxonomy, which are more familiar to them. The adoption of these two systems by China will facilitate academic exchanges, international cooperation, and higher education. For that purpose, we plan to establish different versions of the soil database based on all three soil classification systems. The process will involve establishing links between the soil reference information in nearly 3000 soil profiles (described above) and the digital soil map of China, thereby creating versions of the digital soil map in each of the three systems. These different versions of the digital soil maps can also be linked to the soil profile attributes, eventually forming four different versions of the database: versions based on the "Chinese Soil Genesis Classification System", on the "Chinese Soil Taxonomy", on the USDA "Soil Taxonomy", and on the WRB taxonomy. This will create a convenient tool for boosting international academic exchanges and cooperation and promoting higher education in countries around the world.

6. Conclusions

To date, printed maps and books constitute the sole description of China's soil resources. The development of digital soil maps and associated databases of soil attributes will be a critical step required for rigorous management of China's soil resources. The creation of a reference system for China's soils and versions of the database based on different soil-classification systems will also improve international understanding of Chinese soils and promote broader application of the data.

Acknowledgments

The research was funded by the Natural Science Foundation of Jiangsu Province (Project # BK2002504), an Innovation project of the Chinese Academy of Sciences (ISSASIP0201), and the National Key Basic Research Support Foundation (Project # G 1999011810).

References

FAO (1998) World Reference Base for Soil Resources. Food and Agriculture Organization, Rome. World Soil Resources Reports: 1-87

Gong, Z.T. et al. (1978) A drafting proposal for soil classification of China. *Soil* 5: 168-169

The Officer for the Second National Soil Survey of China (1996) Soil Series of China (in Chinese). Agriculture Press: 1-898

Shi, X. (1998) Soil Geography and GIS technology. Proceedings of the Workshop for Sustainable Use of Land Resources and GIS technology (in Chinese). China GeoScience Press: 23-28

Soil Survey Staff (1994) Keys to Soil Taxonomy. U.S. Department of Agriculture, Washington D.C., Sixth Edition: 1-306

Xi, C.F. (1998) Soils of China (in Chinese; title translated by the authors), Agriculture Press: 1-1253

Construction and use of a soil inventory by NIAES, Japan

Hiroshi Obara, Toshiaki Ohkura, Kazuki Togami and Makoto Nakai

National Institute for Agro-Environmental Sciences, 3-1-3 Kannondai, Tsukuba, Ibaraki 305-8604, Japan

Abstract

Soil inventories provide important basic information to support agricultural and environmental research. The National Institute for Agro-Environmental Sciences (NIAES) soil inventory includes a range of soil information. The database created from this inventory contains analytical data and soil maps produced by the "fundamental soil survey" of arable land across Japan. Soil maps were printed until the 1980s, but were subsequently digitized in the late 1980s. Since the fundamental soil survey was completed, monitoring of arable lands has been carried out to detect changes in soil characteristics. Soil classification systems are also part of the soil inventory, and have been developed simultaneously with the soil survey. Soil samples are also part of the inventory, and soil monoliths from representative soil profiles in Japan have been collected in the NIAES soil museum. The soil inventory data have been used in many studies of agricultural productivity and agro-environmental assessments.

Keywords : soil classification, soil database, soil inventory, soil monolith

1. Introduction

A conceptual framework for the National Institute for Agro-Environmental Sciences (NIAES) soil inventory was first proposed in 2001. This framework aims to integrate various sources of soil-related information (soil classification, soil monoliths, and soil samples; Fig. 1) and promote their use in research.

The first nationwide soil survey was started about 100 years ago in Japan. Since then, various soil surveys have been performed using a variety of survey methods and classification methods. On the basis on the survey results, 1:50 000-scale soil maps of arable land that covered all of Japan were published until 1978. Thereafter, soil maps of arable land began to be digitized (Fig. 2), and 1:50 000 digital soil maps were published on CD-ROM in 2002. The characteristics of arable soils have been monitored each 5 years since 1979 at about 20 000 fixed points, and the results have been entered into a monitoring database. A soil

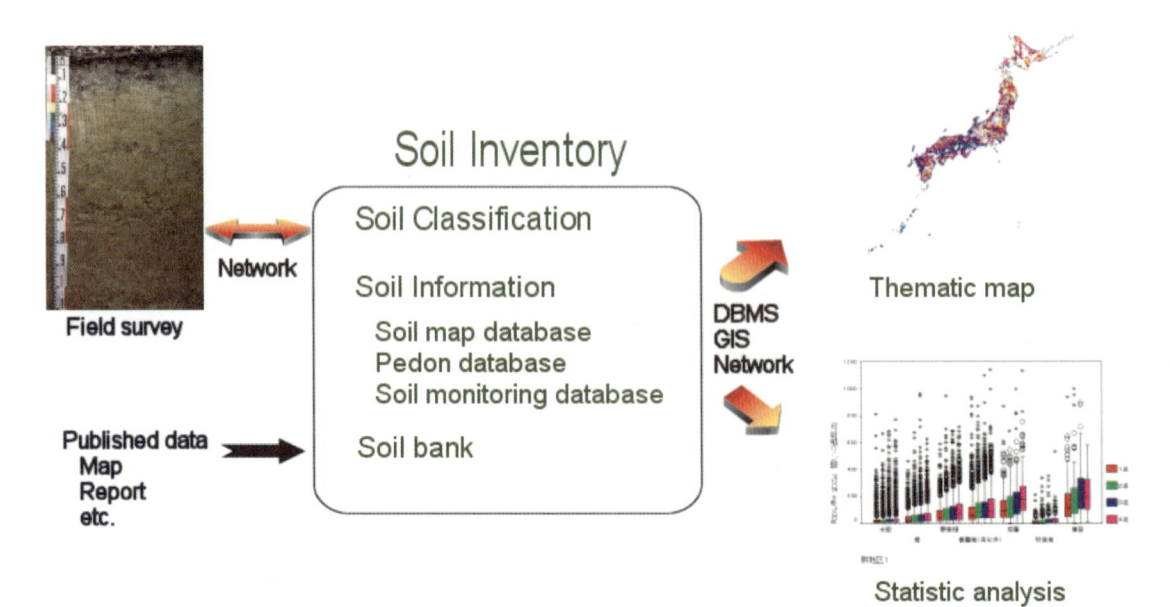

Fig. 1 Image of Soil Inventory

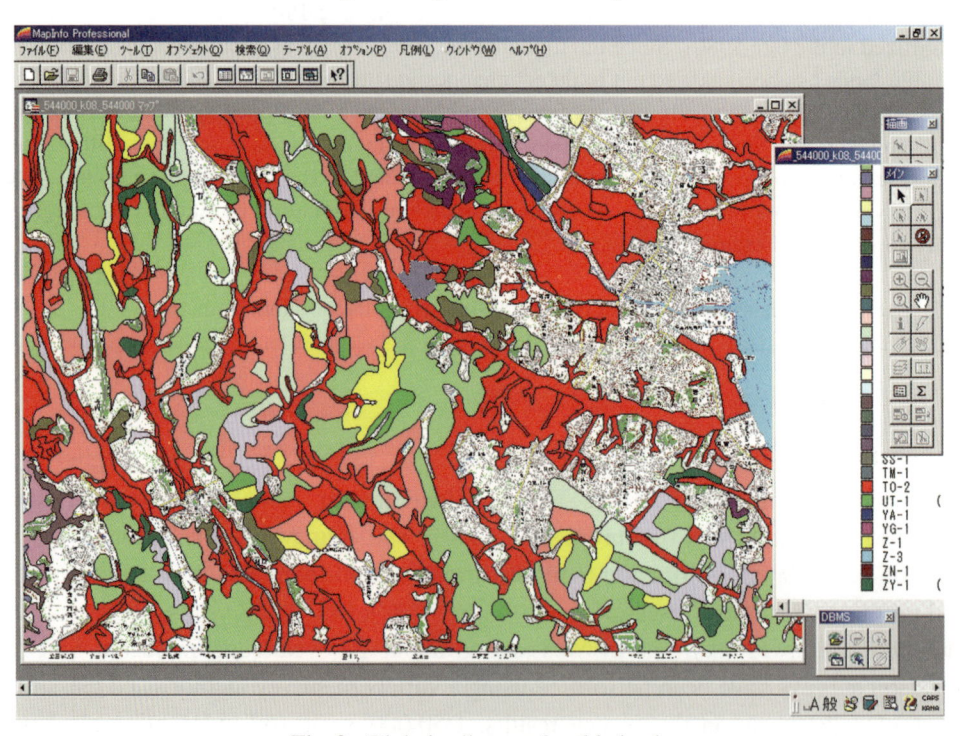

Fig. 2 Digital soil map of arable land

classification system has also been developed on the basis of the data from soil surveys performed since the 1950s. The first approximation of the soil classification system for cultivated land was published in 1973, and second and third approximation were published in 1983 and 1994, respectively. In addition, soil monoliths from typical soils in Japan have been accumulated since 1974 at NIAES. These monoliths contain samples from each soil horizon, and have been preserved for use as environmental standards.

To maintain the valuable resources obtained from fundamental soil research activities and to find ways to better utilize them, NIAES began the construction of a computerized soil inventory system.

2. Japan's soil inventory

2.1. Soil classification

A soil classification system represents the sum of all knowledge concerning the soil resources covered by the system. It is thus the backbone of the inventory data. In Japan, several versions of a soil classification system for arable land have been developed. The second (revised) approximation of the classification system was developed to support the soil classification presented in the 1:50 000 soil maps in 1983. As more knowledge of domestic soils was gained and as international soil classification systems were developed and refined, the Japanese system was significantly revised in 1995. These Japanese classification systems were developed for domestic use, and strictly for cultivated soils. Thus, the current system must be used cautiously for watersheds that include urban and forested areas. The classification system also differs from international soil classification systems such as the World Reference Base for Soil Resources published by the United Nations' Food and Agriculture Organization (FAO) and the United States Department of Agriculture (USDA) Soil Taxonomy. Researchers are currently investigating how to correlate the domestic soil classification system with the international systems.

2.2 Soil databases

2.2.1. Soil survey program for the conservation of soil fertility

Soil maps are basic components of any soil inventory. Japan's soil survey program for soil fertility conservation (1959–1978) produced 1:50 000-scale soil maps that cover all arable land in Japan. This represents the most comprehensive survey of arable land in Japan. The soil maps produced by this program were digitized to create a database containing point information (Kato, 1984). The digital soil maps were issued in 2002 on a CD-ROM. The contents of the database are as follows:

 a) A database of soil maps (1:50 000) for all cultivated soils
 b) Data on 20 431 pedons (Fig. 3)
 c) 69 909 data points that describe all soil layers in the database
 d) 55 561 data points that represent the results of analyses for each layer.

Major components of the NIAES soil inventory are listed in Table 1.

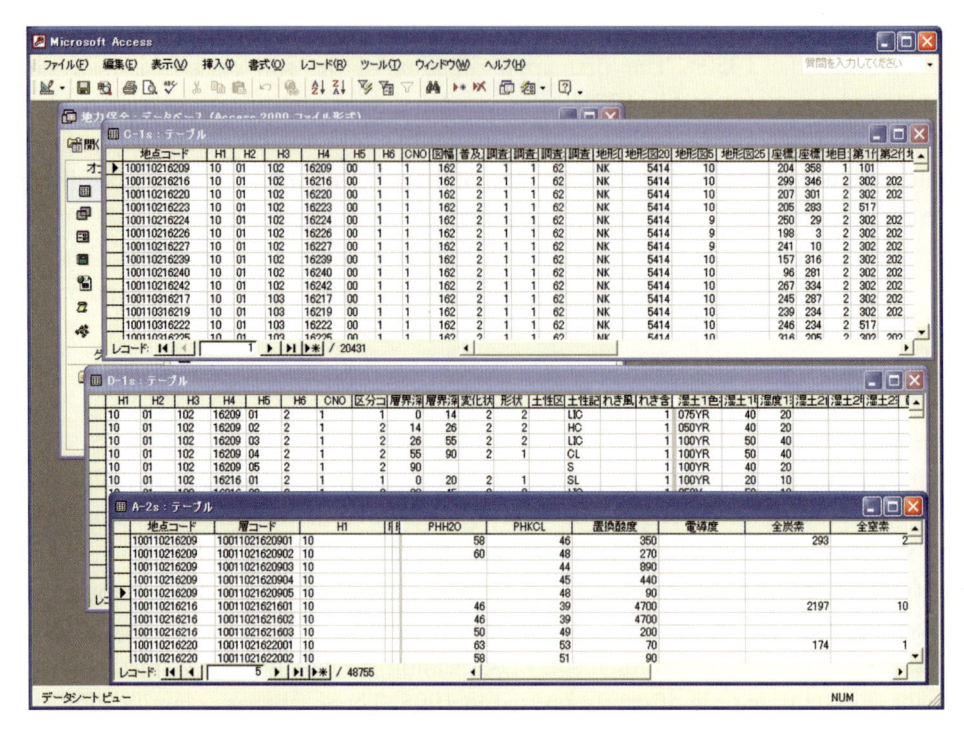

Fig. 3 Pedon database of Soil survey program for soil fertility conservation (MS Access)

Table 1 Estimated major components of Soil Inventory in NIAES

Soil Classification
    Classification of cultivated soil, second approximation edition (1983).
    Classification of cultivated soils in Japan, Third approximation (1995).
    A functional soil classification system for environmental resources. (just started)
Soil Database
    Soil survey program for soil fertility conservation, soil map 1:50,000 database (2002)
    Soil servey program for soil fertility conservation, site database (2002)
    Soil monitoring survey database (under construction)
    Standard soil profile databese (under construction)
    Regional and national soil map database (under construction)
Soil Bank
    Soil monolith harizon sample (626 sample are available)

2.2.2. Soil monitoring program

The characteristics of the arable soils have been monitored each 5 years since 1979 in a network of about 20 000 fixed points (Fig. 4), and the results have been compiled in a database that tracks changes in these values. Various characteristics are monitored at each location:
    1) soil properties (chemical, physical, nutrient, morphology, etc.)
    2) agricultural practices (crop, yield, amounts of fertilizer and manure applied, etc.).

Fig. 4 Location of monitoring sites

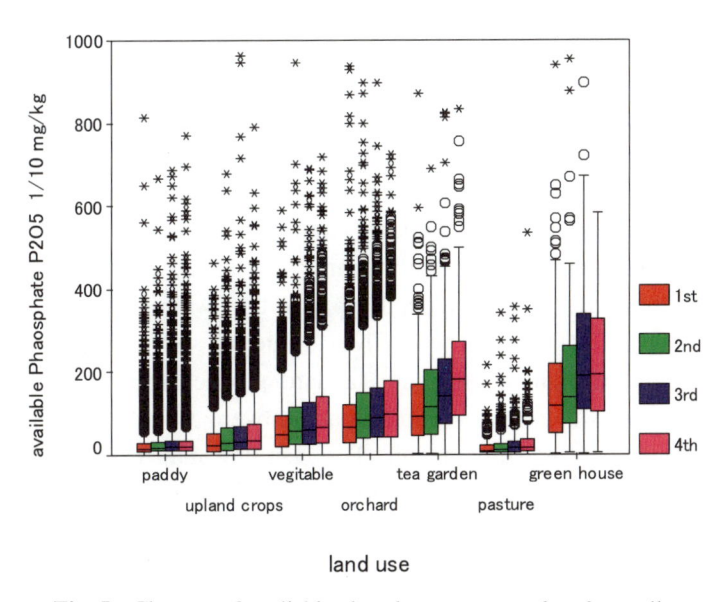

Fig. 5 Changes of available phosphate contents of surface soils

The sites in this database cover the majority of arable land in Japan. Figure 5 shows an example of one type of data that can be visualized by using the database: changes in the available phosphate content of surface soils for several types of crop. Levels of available phosphate clearly increased between 1979 and 1997. In addition to the national soil database, NIAES is developing a database of research results. Since 1974, NIAES has been collecting soil monoliths, and nearly 200 are now stored in the NIAES Soil Museum. We are trying to develop a standard (benchmark) database for soil classification based on the data contained in our soil monolith database.

2.3 Soil bank program

A soil bank represents a collection of soil samples. The collection of soil monoliths for typical soils in Japan developed by NIAES since 1974 is such a soil bank. Samples of each soil horizon have been preserved to serve as environmental standard soil samples (i.e., benchmarks for soil conditions in the year of the sample collection). In addition to these soil samples, we are collecting various soil samples along with corresponding site data and profile descriptions. These samples have been reserved for future study, such as when a researcher wants to be able to calculate changes in levels of soil pollution over time.

3. Use of the soil inventory

After the construction of the first soil information system for arable land (Kato, 1984), many studies used the system to study agricultural productivities and perform agro-environmental assessments. Recent major uses of the soil inventory are listed in Table 2. In this section, we introduce two examples of recent uses of the soil survey data and the preserved soil samples.

3.1 Water conservation by soils in a small catchment

In this example, the database of large-scale soil maps was used to estimate the water conservation capacity of a small catchment. In the 1990s, a study was conducted to develop monitoring methods for the agricultural environments of a small catchment in Yasato Town, Ibaraki Prefecture. As a part of the project, large-scale soil mapping was performed to estimate the water conservation capacity of the area's soils, and a soil map was prepared at a scale of 1:5000. Soils were studied to a depth equal to the level of the water table or bedrock (a maximum of 5 m in depth). Soils were divided into 45 soil groupings based on combinations of soil layers and soil depth, and standard profiles were recorded for each soil grouping. Many kinds of water conservation capacities were estimated by applying layer data to standard profiles for each mapping unit. The resulting map (Fig. 6) shows the water-holding capacity of the area's soils (at water potentials of 0 to 50 kPa).

Table 2 Examples of utilization of soil databases and soil bank

Soil information
    Permissible amount presumption of organic matter input in upland field soil
    Water conservation function of soil in small watershed
    Estimation of cold weather damage of paddy rice
    Arable land soil information system (Chiba prefecture)
    Soil survey support system (Mie prefecture)
Soil bank
    Soil resources and pollution studies at NIAES and another institute

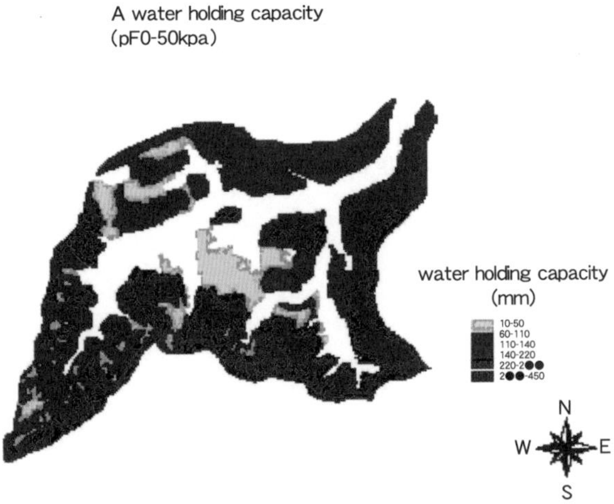

A water holding capacity
(pF0-50kpa)

water holding capacity
(mm)
10-50
60-110
110-140
140-220
220-2●●
2●●-450

Fig. 6 Map of water holding capacity

3.2 Background levels of elements in the soils of Japan

In this example, the soil bank was utilized to clarify the background levels of various elements in soils. The cooperative study involved Tohoku University and NIAES. The soil monoliths in the NIAES collection were used to provide representative data for a wide range of the soil types common in Japan. By using these samples, the concentrations of 57 elements were measured using the latest analytical methods. The results of this analysis were stored in the soil monolith database.

4. Conclusions: ongoing activities

The soil inventory will continue to be used for agro-environmental research. For example, it will respond directly to investigations on soil quality and soil resources. The soil inventory will also play a key role as the source of fundamental data for modeling and for other, more complicated problems. To improve the effectiveness of the soil inventory, the following activities are under way:

- Developing an online soil information system
- Updating the soil monitoring database and analyzing trends in the soil properties of cultivated land
- Improving the soil classification system and correlating it with international systems
- Developing a database on the deep soil characteristics of large watersheds
- Developing a database on levels of heavy metals in soils.

References

Cultivated Soil Classification Committee (1995) Classification of cultivated soils in Japan, Third Approximation. Miscellaneous publication of the National Institute of Agro-Environmental Sciences, Japan. No. 17: 79 pp

Hamazaki, T. et al. (2002) Distribution of soil groupings and water conservation capacities of soils in the Hosaki small watershed. *Jpn. J. Soil Sci. Plant Nutr*. 73: 279−285

Kato, Y. (1984) A computerized soil information system for arable land in Japan, I. Concept, Objective, Process, and Structure. *Soil Sci. Plant Nutr*. 30(3): 287−297

Yamazaki, S., et al (1999) Background levels of trace and ultra-trace elements in soils of Japan. *Soil Sci. Plant Nutr*. 47(4): 755−7654

Construction of an insect inventory and its utilization by Japan's NIAES

Koji Yasuda, Shin-ichi Yoshimatsu and Yukinobu Nakatani

National Institute for Agro-Environmental Sciences, 3-1-3 Kannondai,
Tsukuba, Ibaraki 305-8604, Japan

Abstract

The insect museum of Japan's National Institute of Agro-Environmental Science (NIAES) holds many insect specimens, including dry-pinned, slide-mounted, and alcohol-preserved specimens. These represent donated collections, type specimens, identification requests, and voucher specimens. These specimens are an essential resource for taxonomic research, and their label data, such as the collection location, date, and host record, also provide useful information for environmental and biodiversity research. To facilitate the use of such a vast collection of entomological information, we plan to construct an insect inventory that gathers together the available information and makes it easily accessible through the Web. Some databases are listed as examples of the components of the inventory. A database of specimens collected at a given location would help researchers describe the environment of an area. Moreover, the accumulation of useful information for insect identification might promote the development of an identification-support system that permits the rapid detection of invasive insects or facilitates the analysis of biodiversity data.

Keywords : insect specimens, holotype, identification, environment, biodiversity

1. Introduction

The insect museum of Japan's National Institute of Agro-Environmental Science (NIAES) holds many insect specimens, which have been collected since 1899, when the insect research section of NIAES was established (Matsumura, 1996). These specimens are essential resources for taxonomic research, including the identification and classification of insects. Moreover, every specimen contains other useful information such as the location where it was collected, the collection date, and the host. This information can help to document the environment of the area where the insect was collected.

NIAES is currently working on a project to construct an insect inventory that gathers together a wide range of information on insects and that makes this information more easily accessible through the Web. In addition, we hope that our integration of various kinds of information may result in a new tool that will prove useful for biodiversity research. In this paper, we introduce the insect museum of NIAES and its holdings, and provide examples of how the insect inventory has been used.

2. The insect museum of NIAES

The insect museum (Fig. 1), built in 1979 in Tsukuba, is maintained and developed by the Insect Systematics Laboratory of NIAES. It occupies 600 m^2 of floor space and houses 250 steel cabinets with a total capacity of 12 000 specimen-storage boxes. Currently, 7000 boxes contain dry-pinned specimens (Fig. 2). Some groups of small insects, such as aphids, thrips, and minute parasitoid wasps, have been mounted on slides to permit microscopic observation. More than 600 slide boxes, each of which stores 100 slides, are kept in the museum. Soft-bodied insects, such as the larvae of moths and flies and other miscellaneous small insects caught in traps, are preserved in alcohol. The alcohol-preserved specimens are contained in nearly 4000 glass tubes or plastic containers. In addition to these specimens, there are many unmounted insects, most of which are wrapped in paraffin paper. We estimate the total number of all kinds of specimens to be about 1 200 000.

2.1. Collections of insect specimens

The specimens in the museum mainly consist of collections by successive staff of our laboratory and donations by researchers from other institutes and the general public (Table 1). Each collection has its own special value as a research material because of its concentration on a particular taxonomic group or its collection in a particular area. Some of the collections have had their specimens identified and arranged to improve their usefulness for taxonomic research and the identification of insects.

Fig. 1 Insect museum

2.2. Holotype specimens

A holotype specimen is a single specimen upon which a new species-group taxon is based, and provides the basis for the original publication describing the species. We estimate that nearly 1000 holotypes exist in the museum, though the actual number is not known. Among

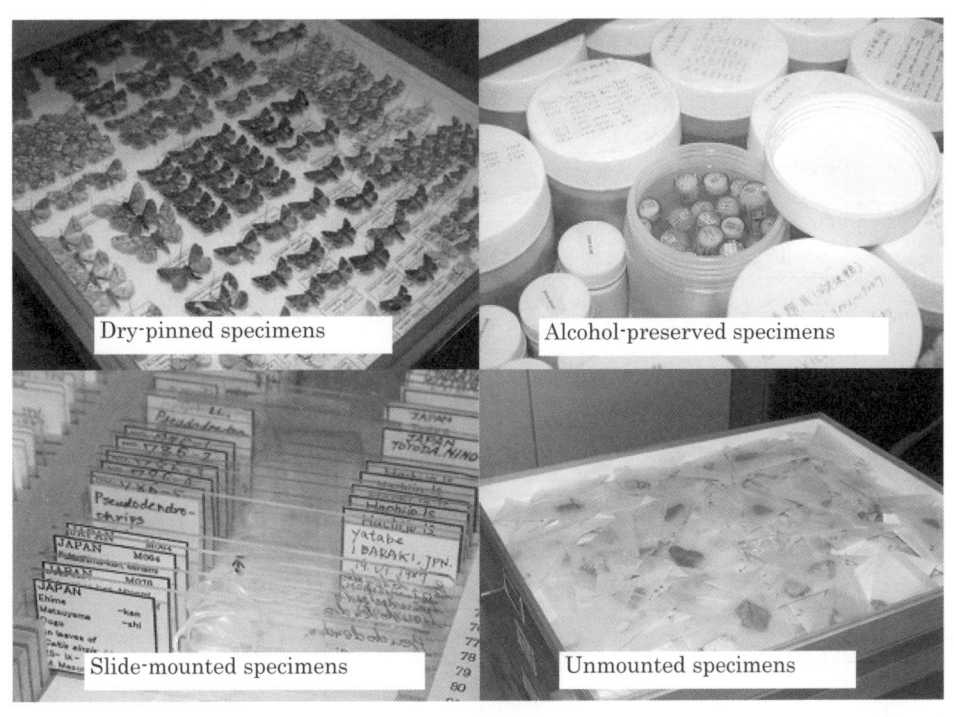

Fig. 2 Various kinds of specimens

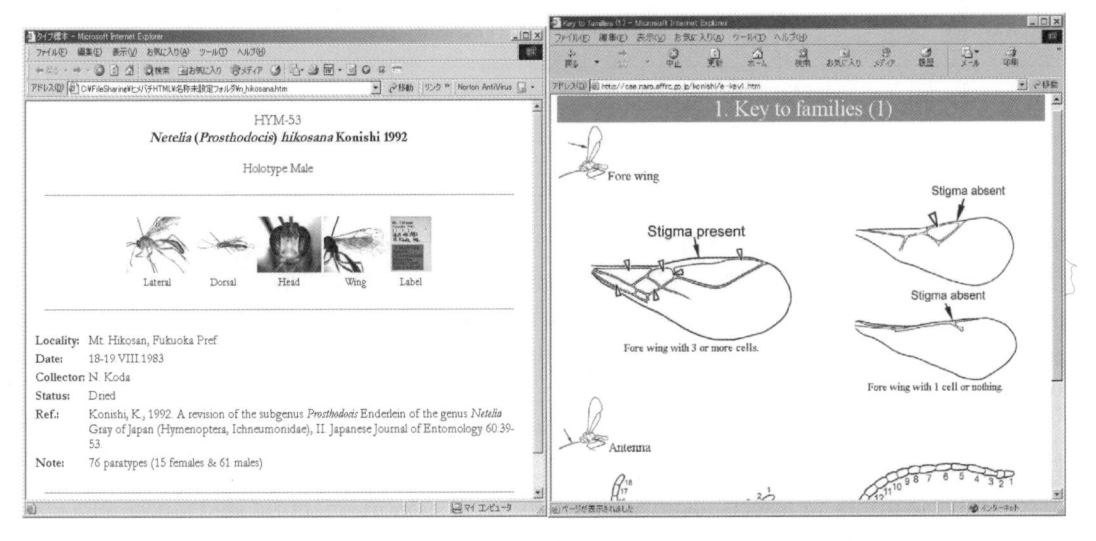

Fig. 3 Holotype database

Fig. 4 Illustrated key to the Hymenopterous parasitoids of *Lirimyza trifolii*

Table 1 Major collections in the insect museum, NIAES

Donor	Contents	Donor	Contents
T. Fujimura	Specimens of Cerambycidae and other beetles from Japan and Southeast Asia	I. Kuwana	5000 specimens of scale insect including types
N. Fukuhara	Dipteran and Orthopteran specimens	M. Kurosawa	5000 slides of Thysanoptera and 61 types
R. Sato	8400 lepidopteran specimens, mainly Noctuidae	S. Kuwayama	3200 specimens of Neuroptera, Mecoptera and other insects collected from Sourthern Kryl Islnds (52 types)
A. Habu	Carabid and chalcidid specimens including 340 types	H. Nakajima	18000 lepidopteran specimens, mainly Geometoridae and Sphingidae
T. Shiraki	Syrphid, tephritid and tabanid specimens including types	Y. Niijima	Thousands number of coleopteran specimens, mainly Scolytidae
I. Hattori	Liquid specimens of lepidopteran larvae	A. Nobuchi	500,000 scolytid spcecimens including types
M. Terayama	Type series of Formicidae	T. Okazaki	Specimens collected from Tokyo
T. Ishii	Parasitoid wasps and Southeastern Insects	N. Oho	Beautiful insects of the world, mainly beetles and butterflies
H. Inoue	Geometorid and Pyralid specimens	H. Furukawa	Orthopteran specimens
S. Kato	Dipteran specimens	H. Hasegawa	Heteropteran specimens
S. Katsuya	17500 specimens, mainly parasitoid wasps	S. Sugi	29000 lepidopteran specimens including types of Noctuidae and Notodontidae
A. Kawada	Almost 10,000 adults and larvae specimens of Lepidoptera	H. Takahashi	Simuliid and tabanid specimens, and related literature
S. Kumasawa	Specimens from Northern Japanese Alps Mts.	H. Hayakawa	Japanese tabanid specimens including types
T. Kumazawa	5000 coleopteran specimens	K. Tsuneki	20,000 formicid specimens

these specimens, 508 (Table 2) are kept in a type specimen room, and the associated data (location of the collection, date, collector's name, and the original literature that describes them) have been compiled into a computer database. Moreover, digital images of the holotypes have been steadily added to the database, which has been partially published on the Web (Figs. 3, 4).

2.3. Identification services

NIAES also offers an insect-identification service to support scientific study, pest management, and the protection of human health, among other important tasks. Many requests for identification are received every year from other agencies and companies in Japan. When we

Table 2 Holotypes in NIAES (2003)

Order	No. of holotypes
Coleoptera (beetle, weevil)	237
Hymenoptera (wasp, ant)	83
Diptera (fly, mosquito)	83
Hemiptera (aphid, sting bug)	59
Thysanoptera (thrips)	24
Neuroptera (race wing)	10
Lepidoptera (butterfly, moth)	9
Trichoptera (caddish fly)	3
Total	508

examine samples for identification, specimens held in the museum are often used as reference materials. The samples are mostly added to the museum collection after identification to serve as a reference for future identification services. Approximately 9000 records of such identifications have been stored since 1948. These records have added a large amount of good information to the insect inventory.

2.4. Voucher specimens

Voucher specimens are used to confirm the original identification of organisms in a study. When questions arise about the identity of these organisms, voucher specimens can be examined to confirm the validity of the original identification. Voucher specimens should be deposited in a public museum to ensure long-term maintenance and availability. The NIAES insect museum accepts requests from other agencies for the deposition of voucher specimens, but such cases are relatively uncommon, because the importance of voucher specimen is unfortunately not well recognized in Japan.

3. Insect inventory

Vast quantities of information and knowledge on insects that have been produced by numerous studies and investigations are occasionally stored separately at a range of locations, without adequate accessibility to researchers. Collecting such information into single, easily accessible databases (e.g., accessible through the Web) is essential for the effective use of this information. Therefore, we have begun a project to construct an online insect-inventory database that links many existing databases. First, we list some databases to provide examples of the components of the inventory. The bottom four are already available through the Web, and the other databases are under construction.

(1) NIAES insect specimen database
(2) NIAES insect holotype database (Fig. 3)
(3) Identification request database
(4) Insect literature database kept in the insect museum of NIAES
(5) Database of invasive insect in agro-ecosystems

(6) Colonies of insects and mites maintained in entomology laboratories of Japan
(7) The low development threshold temperature and the thermal constant in insects, mites, and nematodes in Japan
(8) A checklist of Japanese *Cinara* Curtis (Homoptera: Aphididae) with keys to the species
(9) An illustrated key to the Hymenopterous parasitoids of *Liriomyza trifolii* in Japan (Fig. 4)

4. Utilization of specimen information

Every identified specimen bears information on the habitat conditions under which the specimen was collected. The concentration of such information from numerous identified specimens collected from a particular area can help document the environmental conditions or biodiversity of that area (Fig. 5). Moreover, a series of such information from different dates may reveal changes in the conditions at that location. Thus, the database of identified specimens will contribute greatly to research projects on the environment and on biodiversity.

Moreover, the extensive growth of insect specimen databases will activate taxonomic studies and facilitate the accumulation of information that is useful for insect identification. This data may lead to the development of an effective identification-support system (Fig. 6) that can facilitate the identification of insects. Such a system will speed up the detection of invasive insects and the analysis of biodiversity. To create the identification-support system, the establishment of an international network for sharing data between many museums and institutions will be indispensable.

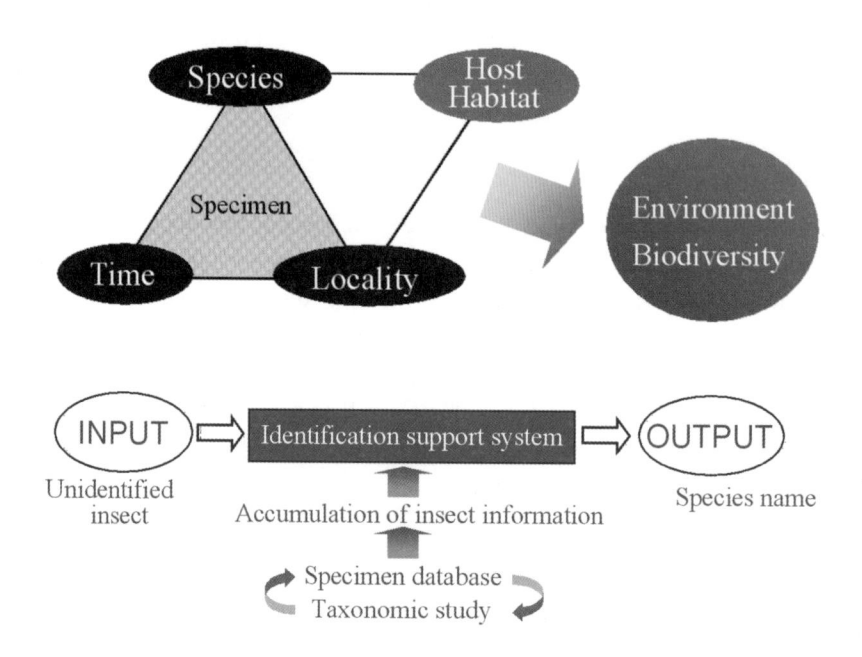

Acknowledgments

We thank our colleagues at the Natural Resources Inventory Center of NIAES for their fruitful discussions on the concept of the insect inventory.

References

Matsumura, T. (1996) Insect specimens as agro-environment resources. Explanation Series No. 3, National Institute of Agro-environmental Sciences, Japan: 18 pp

2005	2005年2月25日 第1版発行	
農環研英文叢書 第5号 東アジアの農業 生態系における 物質循環と環境 影響評価		
検印省略	著 作 者	茨城県つくば市観音台3-1-3 独立行政法人農業環境技術研究所
©著作権所有	発 行 者	株式会社 養 賢 堂 代表者 及 川 清
定価 21,000 円 (本体 20,000 円) (税　5％)	印 刷 者	株式会社 丸井工文社 責任者 今井晋太郎
発 行 所	〒113-0033 東京都文京区本郷5丁目30番15号 株式会社 養 賢 堂　TEL 東京(03)3814-0911 振替00120 FAX 東京(03)3812-2615 7-25700 URL http://www.yokendo.com/	
	ISBN4-8425-0368-8 C3061	
PRINTED IN JAPAN	製本所　株式会社丸井工文社	